Undergraduate Texts in Mathematics

Editors

S. Axler
F.W. Gehring
P.R. Halmos

Springer

New York
Berlin
Heidelberg
Barcelona
Budapest
Hong Kong
London
Milan
Paris
Santa Clara
Singapore
Tokyo

Undergraduate Texts in Mathematics

Anglin: Mathematics: A Concise History and Philosophy.
Readings in Mathematics.

Anglin/Lambek: The Heritage of Thales.
Readings in Mathematics.

Apostol: Introduction to Analytic Number Theory. Second edition.

Armstrong: Basic Topology.

Armstrong: Groups and Symmetry.

Axler: Linear Algebra Done Right.

Bak/Newman: Complex Analysis.

Banchoff/Wermer: Linear Algebra Through Geometry. Second edition.

Berberian: A First Course in Real Analysis.

Brémaud: An Introduction to Probabilistic Modeling.

Bressoud: Factorization and Primality Testing.

Bressoud: Second Year Calculus.
Readings in Mathematics.

Brickman: Mathematical Introduction to Linear Programming and Game Theory.

Browder: Mathematical Analysis: An Introduction.

Cederberg: A Course in Modern Geometries.

Childs: A Concrete Introduction to Higher Algebra. Second edition.

Chung: Elementary Probability Theory with Stochastic Processes. Third edition.

Cox/Little/O'Shea: Ideals, Varieties, and Algorithms.

Croom: Basic Concepts of Algebraic Topology.

Curtis: Linear Algebra: An Introductory Approach. Fourth edition.

Devlin: The Joy of Sets: Fundamentals of Contemporary Set Theory. Second edition.

Dixmier: General Topology.

Driver: Why Math?

Ebbinghaus/Flum/Thomas: Mathematical Logic. Second edition.

Edgar: Measure, Topology, and Fractal Geometry.

Elaydi: Introduction to Difference Equations.

Exner: An Accompaniment to Higher Mathematics.

Fischer: Intermediate Real Analysis.

Flanigan/Kazdan: Calculus Two: Linear and Nonlinear Functions. Second edition.

Fleming: Functions of Several Variables. Second edition.

Foulds: Combinatorial Optimization for Undergraduates.

Foulds: Optimization Techniques: An Introduction.

Franklin: Methods of Mathematical Economics.

Hairer/Wanner: Analysis by Its History.
Readings in Mathematics.

Halmos: Finite-Dimensional Vector Spaces. Second edition.

Halmos: Naive Set Theory.

Hämmerlin/Hoffmann: Numerical Mathematics.
Readings in Mathematics.

Iooss/Joseph: Elementary Stability and Bifurcation Theory. Second edition.

Isaac: The Pleasures of Probability.
Readings in Mathematics.

James: Topological and Uniform Spaces.

Jänich: Linear Algebra.

Jänich: Topology.

Kemeny/Snell: Finite Markov Chains.

Kinsey: Topology of Surfaces.

Klambauer: Aspects of Calculus.

Lang: A First Course in Calculus. Fifth edition.

Lang: Calculus of Several Variables. Third edition.

Lang: Introduction to Linear Algebra. Second edition.

Lang: Linear Algebra. Third edition.

Lang: Undergraduate Algebra. Second edition.

Lang: Undergraduate Analysis.

Lax/Burstein/Lax: Calculus with Applications and Computing. Volume 1.

(continued after index)

George R. Exner

An Accompaniment
to Higher Mathematics

Springer

George R. Exner
Department of Mathematics
Bucknell University
Lewisburg, PA 17837
USA

With nine illustrations

Mathematics Subject Classification (1991): 00A05

Library of Congress Cataloging-in-Publication Data
Exner, George R.
    An accompaniment to higher mathematics / George R. Exner.
        p.   cm. — (Undergraduate texts in mathematics)
    Includes bibliographical references and index.
    ISBN 0-387-94617-9 (hardcover : alk. paper)
    1. Proof theory. I. Title. II. Series.
QA9.54.E96   1996
511.3 — dc20   95-44884

Printed on acid-free paper.

Production managed by Frank Ganz; manufacturing supervised by Jeffrey Taub.
Photocomposed pages prepared from the author's LaTeX files.
Printed and bound by R.R. Donnelley and Sons, Harrisonburg, VA.
Printed in the United States of America.

9 8 7 6 5 4 3 2 1

ISBN 0-387-94617-9 Springer-Verlag New York Berlin Heidelberg

If ideas came quite at random, the solution of problems would depend mainly on chance. Many people believe that this is so. ... It is difficult to believe that such a widespread opinion should be entirely devoid of foundation, completely wrong. But is it completely right?

George Pólya, *Mathematical Discovery: On Understanding, Learning, and Teaching Problem Solving (combined edition)* (Ref. [1]), ©1981 by John Wiley & Sons, Inc. Reprinted by permission of John Wiley & Sons, Inc.

To many, mathematics is a collection of theorems. For me, mathematics is a collection of examples; a theorem is a statement about a collection of examples and the purpose of proving theorems is to classify and explain the examples ...

John B. Conway, *Subnormal Operators* (Ref. [2]), Research Notes in Mathematics No. 51, Pitman Advanced Publishing Program, 1981. Used by permission.

"When *I* use a word," Humpty Dumpty said, in a rather scornful tone, "it means just what I choose it to mean — neither more nor less."

Lewis Carroll, *Through the Looking Glass, and What Alice Found There* (Ref. [3])

In memory of my father
Robert M. Exner

Contents

Introduction

For Students

Congratulations! You are about to take a course in mathematical proof. If you are nervous about the whole thing, this book is for you (if not, please read the second and third paragraphs in the introduction for professors following this, so you won't feel left out). The rumors are true; a first course in proof may be very hard because you will have to do three things that are probably new to you:

1. Read mathematics independently.

2. Understand proofs on your own.

3. Discover and write your own proofs.

This book is all about what to do if this list is threatening because you "never read your calculus book" or "can't do proofs." Here's the good news: you must be good at mathematics or you wouldn't have gotten this far. Here's the bad news: what worked before may not work this time. Success may lie in improving or discarding many habits that were good enough once but aren't now. Let's see how we've gotten to a point at which someone could dare to imply that you have bad habits.[1]

The typical elementary and high school mathematics education in the United States tends to teach students to have ineffective learning habits,

---

[1] In the first paragraph, yet.

and we blush to admit college can be just as bad. A lecture designed to make it unnecessary to read the book, for example, tends to make you not read the book. If lecture and text present you with a list of formulas to memorize and model problems to imitate, you reasonably assume mathematics consists of memorization and imitation. The answers in the back of the book teach you that the answer is the important thing. If the book is written so that there is no place for a reader to participate actively (by asking questions, constructing examples, or solving problems), you are trained to believe that reading mathematics is a passive activity. If you are not given effective tools for word problems (story problems), you use ineffective methods of attack (two popular favorites are frantic but unthoughtful activity and an endeavor to keep the problem at least 10 feet away and poke it with a stick to see if it is safely dead). If the benefits of good notational habits are not shown, you adopt poor notational habits [for example, it is frequent in the taking of derivatives to write, but not to mean, that $f(x) = x^2 = 2x$]. If a test consists only of problems exactly like those in the book, you are encouraged to believe that doing mathematics is having success at doing old problems.

To teach you this sort of thing is to do you an injury. The behaviors that result are learned, as is clear because children have active and effective learning habits (as anyone who has watched a five-year-old take something apart can verify). As adults, we solve problems every day, and wouldn't dream of being passive, purely imitative, or only answer oriented in non mathematical areas. Those of you in athletics, for example, would feel cheated if your sports season (the chance to play) was replaced by just the final scores of all the games (the "answer"). Likewise, the pleasure in music is primarily the doing, and the struggle to learn and perfect new music, not endless time playing old music over and over.

Like music and athletics, the main pleasure in mathematics is in the process of doing: exploring, inventing, and discovering. Again, the good news is that somehow you survived intact enough to make it to a first course in proof. You are therefore intrinsically *good* doers of mathematics, but you may be fighting with one hand (or two hands) tied behind your backs without knowing it. This book is all about some active tools to make your doing of mathematics more effective and thus, possibly, more enjoyable.

Like any other tools, these are only good if you use them. This book will be a waste if you are unwilling to try the things it suggests. Symbolic of the sort of thing required is this icon:

This icon (supposed to be a waiting pad of paper and a pencil) means that at this stage you ought to put down the book and do something.

Really. Every time. To keep reading is to miss the whole point of more effective things to do to help your reading and doing; more effective things *not done* are not effective. Go to it.[2]

For Professors

This book is designed for those students about to engage in their first struggle with reading mathematics independently, understanding proofs, and discovering proofs. Such grappling is typically done in either a first course in real analysis, abstract algebra, or sometimes topology; or in a "transitions" course. The changes are dramatic: "don't prove" (calculus) to "prove"; "don't read, imitate examples" (calculus) to "read and think independently." We as professors tend to hope that somehow students are already at the top of the first of the cliffs along the way to being mathematicians; all too often, they simply run into the bottom of that cliff. For all but a few, calculus and linear algebra have not provided much of a ramp up.

The author, at least, persists in approaching this first course in proof with great pleasure and anticipation.[3] The material is exciting and basic to much of the rest of mathematics, and students finally get to see what mathematics is really about. There is usually a large fund of relevant examples with which to aid the development of intuition, and often even simple examples touch the essence of the subject. The entry-level proofs are straightforward and their key ideas well indicated by any decent example; many of the proofs go farther and "write themselves." There are always marvelous concepts that suggest enough problems to keep any student enthralled. What could possibly go wrong?

Except for the students reading this we all know what goes wrong (and they have plenty of gloomy rumors from their elder peers). Whatever your institution's first course in proof, it may well be viewed by students as a baptism by fire for those interested in theoretical mathematics and a painful, grueling, and unnecessary final requirement for those interested in applied work (and too many students abruptly change to applied areas after the course, too). A great deal of student time is spent floundering around and doing things that are, frankly, completely off track. It is not uncommon to be asked about a problem by a student who can give no concrete example of anything relevant to the matter at hand (sometimes in spite of considerable time "working" on the problem). The first proof course would be perfect if we were willing to frustrate, discourage, and fail many students whose mathematical pasts were both successful and enjoyable.

---

[2]Or, if not, sell this thing back to the bookstore.
[3]This is naive but by no means unique.

This book is intended for students whose backgrounds need repair before they or we can decide how far their talents will take them. It presents a pair of tools for reading and doing mathematics (especially proofs) that are primarily processes for learning independently. The effective use of *examples* and the benefits of sensitivity to *informal and formal language* are part of the working stock of any mathematician. The author believes that a student in a first proof course who has not already learned these skills by osmosis is unlikely to do so now, and thus explicit instruction is required. The goal is to help students become active and independent learners and doers of mathematics by showing them some effective ways to do so.

Ways to Use This Book

This book can serve as a supplement to any of the standard texts for a traditional "content" course in real analysis, abstract algebra, or topology. The most straightforward use is to begin the semester with the chapters about how to read independently and grapple with proof as an investment for the future. This initial investment is nonetheless, for many professors, acutely painful, since it decreases the time for the mathematical content and especially those beautiful theorems that already have to be crowded in at the end. This pain is real, but we've tried to use the standard introductory material (such as sets, functions, relations, and so on) so that this work can be read along with, or even instead of, that standard introduction. One might, for example, assign part of this text, Lab I (Sets by Example) or the equivalent section in the content text, more of this text, Lab II (Functions by Example) or its equivalent, and so on. Further, the exercises are diverse enough so that a Real Analysis or Topology student can be required to construct examples for sequences, for example, while a student in Abstract Algebra can construct instead examples for groups. To some extent, time spent on these techniques doubles as time on introductory content.

This text may also be used as the main text for a "transitions" or "bridge" course. One would begin with the methods chapters, probably interspersed with the Laboratory sections as appropriate. For a full credit course, some supplement near the end of the semester with a content text would probably be required, although students benefit from more time on the methods sections than one would believe possible. A "Moore method" text of some kind provides in many ways the most obvious reinforcement that the techniques from this book are needed (a nice example is the set of notes on Graph Theory written by Martin Lewinter [4]).

Acknowledgments

This book was written over a space of years, and has benefited greatly from comments on its various incarnations from students too numerous to mention, but to whom I wish to give sincere thanks. I would like to thank as well the College of Wooster, which provided technical support during some stages of the production of the manuscript. I wish especially to thank my father, Robert M. Exner, both for insightful comments on an early version of this work and for an inspiring model of thoughtful teaching. Various colleagues, especially those at Bucknell University, also have my gratitude for their encouragement and criticism of this manuscript, as well as their insights into teaching. This approach, and its sins of commission and omission, is my own, but my teaching has been influenced and improved by theirs: thanks. Finally, to Claudia, Cameron, and Laurel, who have paid some of the price for this work's completion: Understand that you make it all worthwhile.

1
Examples

1.1 Propaganda

The argument of this chapter is simple: to understand the abstract ideas of mathematics you need to attach to them concrete examples. People do this all the time in non mathematical areas; think, for example, of a child's brush with the definition of "mammal." According to Webster, a mammal is "one of a class of animals that nourish their young with milk secreted by mammary glands and have the skin usually more or less covered with hair." Rote memorization of this doesn't guarantee much, but if the child can say "Well, a cat is a mammal" the child actually knows something. If the child can run through the checklist and verify it for a cat, things are better yet. If the child knows that a mouse and a whale and a bat are mammals the understanding is pretty good.[1] If the child can say as well that a fish and a bird and a shark are not mammals, and why not, the understanding is really sophisticated. This package of basic examples, examples that display the boundaries of the definition, and "non-examples" is vital; if the child doesn't have it, understanding isn't complete.

You've used examples in mathematics to good effect too. In your calculus courses, for example, you almost certainly solved your first max/min problems by finding an example problem in the text and imitating it as closely as possible. If someone asks you to discuss "lines" you might well say "let's

---

[1] If the child knows a duck-billed platypus is a mammal, the understanding is fairly deep.

take $y = x$ to begin with." You may well have understood the definition of "continuous at a point" by a collection of various functions continuous at a point and, at least as important, functions that exhibit various ways continuity at a point might fail. Your understanding of *subspace* in linear algebra was probably improved when you worked through the details that the collection of polynomials which are zero at $x = 1$ is in fact a subspace of the space of polynomials. Indeed, this paragraph itself is a collection of examples from your probable mathematics histories illustrating the idea of learning by example.

It may not have struck you at the time that the examples you collected were usually supplied by someone else, like a teacher or textbook author. These examples are better than none at all, but, frankly, this is a bad habit we the teachers and you the students have fallen into. When you start doing upper-level college mathematics, or are in graduate or professional work, and certainly when you aren't a student any more and have to learn something on your own, there won't be pre-prepared examples. Part of learning independently is constructing your own examples for your understanding.

The news really isn't bad, since you get more out of constructing your own anyway. You'll understand new things more quickly if you do. An effort to generate your own examples, even if unsuccessful, will lead you to weigh and test each part of a theorem or definition. If you do this regularly, you build a collection of examples you can reuse, and you may see patterns in your collection that lead to good questions. This section is about starting out on the process of using examples in a careful way to help you learn.

Here's a final piece of argument. It is annoyingly trite to say that what you do yourself sticks with you while what you watch doesn't; it is annoying mostly because it is true. The effort to construct an example is an effort, and as such is active instead of passive. You know that *watching* a soccer player doesn't make you a soccer player, and *observing* a fine violinist isn't going to make you first chair in the orchestra. Mathematics is no different.

1.2   Basic Examples for Definitions

We present the technique of constructing basic examples by example, using a definition you might recall from calculus.

Definition 1.2.1 *A function f is <u>injective</u> if, for every x_1 and x_2 in the domain of f, $f(x_1) = f(x_2)$ implies $x_1 = x_2$.*

(You might have used instead the phrase <u>one</u>-<u>to</u>-<u>one</u> for this.)

Unless you are the victim of a conspiracy, there probably are some functions out there that are, in fact, injective. Why don't you pick a function and we'll see if it is?

1.1:

Ahem ... why don't *you* pick a function and we'll see?

1.2:

We can illustrate what needs to be done with f given by $f(x) = x^2$. What we are to check is whether, for every x_1 and x_2 in the domain of f, $f(x_1) = f(x_2)$ implies $x_1 = x_2$. Well, is it true?

1.3:

If the condition seems complicated, realize that one advantage of our concrete example is specificity. We need to check whether, for every x_1 and x_2 in the domain of f, $x_1^2 = x_2^2$ implies $x_1 = x_2$. Does this hold?

1.4:

If this is still too complicated, realize that it is legal (in fact, really the whole point) to be more concrete yet. Let's take $x_1 = 3$. Then $x_1^2 = 9$. We have to check if $x_2^2 = x_1^2 = 9$ implies $x_2 = 3$. Does it?

1.5:

Now we are in business. We have produced an x_1, namely 3, and an x_2, namely ?, such that $x_1^2 = x_2^2$ but $x_1 \neq x_2$.

Perhaps you are complaining (complain, complain, moan, moan, gripe, gripe) that we haven't constructed an example of something satisfying the definition, but of something *not* satisfying the definition. That's quite true. But it is quite likely that you still understand the definition better than you did, and in fact we've constructed a "non-example" (a useful class of things we'll get to later).

Before we turn to another example, let's see what we can get out of this one.[2] Recall that pictures (in this case, graphs) are very useful examples. Draw the graph of the square function and see if you can indicate graphically why it fails to be injective.

---

[2] It is reasonable to hope that your work, even though not completely successful, should get you something. Don't give up on it too soon.

1.6:

As a check, think about the function g given by $g(x) = x^4$. Do you think g is injective? Does the idea of your picture above help in thinking about this? Is g injective?

1.7:

To continue the search for something actually satisfying the definition, consider the function i given by $i(x) = x$. Is this injective?

1.8:

Did however you thought about it at least include a picture?

1.9:

We at last have a basic example for the definition, namely a simple object satisfying its condition. In some ways we'll discuss later, this is a little too special an example. What about g given by $g(x) = 3x + 5$?

1.10:

When done with this, note that the function "x" works, "x^2" does not, and "x^4" does not. *Well?* Ask and answer the obvious question.

1.11:

1.2.1 Exercises

1.12: We warm up with some work on sets. Give some examples of a set described by listing its elements. Give some examples of a set described by a condition (the template for such sets is $\{x : x \ldots\}$, which form is sometimes called <u>set builder notation</u>).

1.13: Suppose that A and B are sets. Recall that the <u>intersection</u> of A and B, denoted $A \cap B$, is the set of all elements in both A and B. Recall that the <u>union</u> of A and B, denoted $A \cup B$, is the set of all elements in either A or B. Define $A - B$ to be the set of all elements in A which are not in B. Examples, please.

1.14: The <u>Cartesian</u> <u>product</u> of the sets A and B, denoted $A \times B$, is the set of all ordered pairs whose first element is in A and whose second element is in B. Find examples; don't start with $A = B$ but do include that special case.

1.15: Here's a definition.

Definition *A function f with range contained in a set S is <u>surjective</u> on S if for every s in S there is an x in the domain of f such that $f(x) = s$.*

You may have met this definition using <u>onto</u> instead of surjective. Explore this definition with various examples and non-examples. It is interesting to keep the same function and change the set S, among other things.

1.16: A function f from A to B is a <u>one</u>-to-<u>one</u> <u>correspondence</u> if f is both injective and surjective on B. Explore.

1.17: A <u>graph</u> is a set of <u>vertices</u> (sometimes called <u>points</u>) and <u>edges</u> (sometimes called <u>lines</u>) connecting some pairs of points. If vertices are joined by an edge, they are <u>adjacent</u>. A <u>walk</u> from vertex v_1 to vertex v_2 is a sequence of vertices beginning with v_1 and ending with v_2, and such that each vertex (except v_1) is adjacent to its predecessor in the sequence. (Note that the edges between the vertices are implicit in the sequence.) A <u>trail</u> is a walk in which no (implicitly traversed) edge is repeated. A <u>path</u> is a walk with no repeated vertices.

It is convenient when considering graphs to draw points and connect them by straight or curved edges; crossings that occur but do not correspond to a vertex don't count. Explore the above definitions.

1.18: Denote the set of edges of a graph G by E_G and the set of vertices by V_G. Two graphs G_1 and G_2 are <u>isomorphic</u> if there is a one-to-one function f from the V_{G_1} onto V_{G_2} such that vertices u and v in V_{G_1} are adjacent if and only if vertices $f(u)$ and $f(v)$ in V_{G_2} are adjacent. Explore these definitions. By the way, the name is again meaningful: "iso" means "same" (e.g., isosceles triangle) and "morphos" is "structure" (biologists might recall the word "morphology"), and isomorphic graphs turn out to have the same structure as far as any question of graph theory is concerned.

1.19: A graph is <u>connected</u> if every two of its vertices have a walk beginning at one and ending at the other. A <u>cycle</u> is a walk in which the first and last vertex are the same, and which is otherwise a path (that is, it includes no other repeated vertices, nor is the first repeated more than once). Explore these.

1.20: A <u>relation</u> from A to B is a subset of $A \times B$, so it is a set of ordered pairs whose first element is in A and whose second element is in B. Give examples using $A = \{1, 2, 3\}$ and $B = \{a, b, c\}$ to start with; what is the difference between $A \times B$ and a relation from A to B?

1.21: A <u>relation</u> <u>on</u> a set S is a relation from S to S. Alternatively, a <u>relation</u> <u>on</u> a set S is defined to be a set of ordered pairs of elements of S. Using $S = \{1, 2, 3, 4\}$ to start with, find some relations on S.

1.22: The definition in the previous problem(s) may have surprised you, since we usually think of a relation as something occurring between elements: "less than" is a relation on the real numbers, for example, and if we substitute into the phrase "$a < b$" some pair of numbers a and b, we get something either true or false. If the result is true we say that a and b satisfy the relation "less than," or just "a is less than b." The connection with the definition above is that if we collect all the pairs of numbers a and b for which we wish "less than" to be true, the set of all these pairs captures the relation "less than" completely. One advantage of calling the set of ordered pairs the relation (instead of calling the "rule" the relation) is that the definition is in terms of things we already know about: sets and ordered pairs. Another advantage is that this way one can get relations very difficult to describe in terms of rules, but which are perfectly good sets of ordered pairs. So from now on we consider the set of ordered pairs the relation, although we are of course happy if you recognize it as something familiar. Using the set S of the previous problem, find the set of pairs for your old friend, what you used to call the relation "less than."

1.23: Find, using the set S of the previous problems, some other familiar relations. Find some other familiar relations in more general settings.

1.24: Relations may have various properties, some of which are important enough to have names. A relation R on a set S is said to be <u>reflexive</u> if for every a in the set S, (a, a) is in R. A relation R is <u>symmetric</u> if whenever $(a, b) \in R$, then also $(b, a) \in R$. A relation R is <u>transitive</u> if whenever (a, b) and (b, c) are in R, then so is (a, c). Explore these definitions on the set S of the previous problems, and in general. Part of your exploration should be to decide why these names are appropriate labels for these various notions.

1.25: A relation which is reflexive, symmetric, and transitive is called an <u>equivalence</u> relation. Find some equivalence relations, and find also some relations that are not. Please note that for a single set S (with more than one element), the plural form "relation<u>s</u>" is appropriate.

1.26: A group is, informally, some set on which we have an operation (which we will call addition and denote "+") with at least some of the familiar properties. Take it on faith for the moment that the real numbers under the operation of addition form a group.[3] A function f from a group G to a

---

[3]They actually have so many extra properties they are much more, but that's a story for another time.

group K is a homomorphism if it preserves addition in the following sense:

$$f(x + y) = f(x) + f(y), \qquad \text{for all } x, y \text{ in } G.$$

Taking each of G and K to be the real numbers, find some examples of homomorphisms. [It should be automatic by now to note in passing that "homo" means "same" and to recognize your recent acquaintance "morphism" (see Exercise 1.18).]

1.27: A set A of numbers is said to be bounded above if there exists some number M so that for every x in A, $x \leq M$. Such a number M is called an upper bound for A. Give examples of sets with an upper bound and some without. If a set has an upper bound, does it have only one?

1.28: A set A that is bounded above is said to have a least upper bound if there is a number M that is an upper bound for A and has the property that no number less than M is an upper bound for A. Explore.

1.3   Basic Examples for Theorems

We may also make basic examples for theorems, but it's worthwhile first to say more carefully what a definition, and then what a theorem, is. The key notation here is that of *condition*; for the moment, think of a condition informally as something that becomes either true or false when we plug in the name of an object (for example, "triangle x has three congruent sides," "x is a set of ordered pairs," ...). A definition is really a name attached to a condition. It is a little complicated because that name may be a noun (a relation is a set of ordered pairs, a triangle is ...) or an adjective (a triangle is equilateral, a function is continuous if ...). The condition may be complex, including several conditions in a bundle, but we agree to use a certain label (noun or adjective, as we're told) for anything satisfying everything in the bundle. Sometimes we say "has a property" instead of "satisfies a condition." In this language the technique we just used to get a basic example for a definition is really a basic example for the *condition*.

Again in this language, many theorems are really a guarantee involving two conditions (we'll worry about other forms later). The hypothesis of the theorem (the assumption) is a condition C_h and the conclusion is another condition C_c. The theorem is the guarantee that any object satisfying the condition C_h necessarily satisfies another condition, namely C_c. The construction of a basic example for a theorem is then really the construction of an example satisfying the condition of the hypothesis and then the verification that the example really does satisfy the conclusion. (Of course, the theorem *already* guarantees it satisfies the conclusion, but the verification is still good for your intuition and your efforts to understand the theorem.)

Let us begin by illustrating the technique with a familiar theorem.

Theorem 1.3.1 *If f is a function continuous on $[a,b]$ and differentiable on (a,b) then there exists a point c in (a,b) such that*

(1.1)
$$f'(c) = \frac{f(b) - f(a)}{b - a}.$$

We shall construct some examples to aid our understanding of this, the Mean Value Theorem.

A starting place is to try to understand what the theorem says by choosing a simple function f and a simple a and b, and trying it out. What about f given by $f(x) = x^2$, $a = 2$, and $b = 5$ as a reasonable starting point?[4] There are several things to check before we proceed further, lest we use a function not fitting the hypotheses of the theorem. For example, the function f is continuous on $[2, 5]$ as needed. What else must be checked?

1.29:

If you didn't check that f was differentiable on (a,b) you need some sort of wakeup call. But realize that what you are doing is getting a simple example for the condition of the hypothesis.

We can compute the right-hand side of equation (1.1):

$$\frac{f(5) - f(2)}{5 - 2} = \frac{25 - 4}{5 - 2} = 7.$$

This is presumably equal to $f'(c)$ for some c in (a, b). Well, for our example $f'(x) = 2x$, so $f'(c) = 2c$. Therefore, if this theorem is correct we should have

$$2c = 7$$

for some c in $(2, 5)$. What is the obvious candidate?

1.30:

(Hey ... the theorem worked!)

There are a few parts of the procedure of constructing this example (as opposed to the example itself) worth noting. We let this sort of thing pass in our construction of an example for a definition, but let's think about it now. Why f given by $f(x) = x^2$?

1.31:

---

[4]Observe that in a better world you would have chosen your own. Since we are just starting out, you get away with it this time.

A more subtle question is, why not use $f(x) = x$?

1.32:

Why 2 and 5?

1.33:

The point is that for this sort of example you should pick one you can compute with as long as it is not misleadingly special (this is the sense in which $i(x) = x$ was *too* special an example for a previous discussion of the definition of injective).

What might you do as an accompaniment to the example above?

1.34:

Remember pictures?[5] The ingrained habit of at least trying to draw a picture gets one *at least* as far as

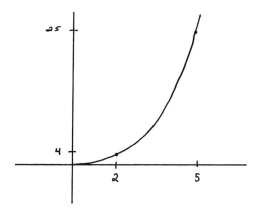

(note that we stick with our concrete example). Even a little understanding of the significance of $f'(c) = f'(7/2)$ should impel you to draw the tangent

---

[5] A cynical teacher might think your eye skipped right to this paragraph without giving you a chance to think. No. You wouldn't do that. Anymore. Please?

line of which it is the slope:

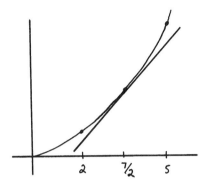

After all, the more of what we know the picture includes, the more useful it is likely to be. The slope $f'(c)$ is equal to

$$\frac{f(5) - f(2)}{5 - 2} = \frac{25 - 4}{5 - 2}$$

so it should be tempting to try to interpret this latter quantity as a slope. Insert a good line with this slope in the diagram!

1.35:

Oh yes, the slope of the chord from $(2, 4)$ to $(5, 25)$ is the right thing, and the picture is complete. Memories of first term calculus might have gotten you here right away; we're claiming you could have done this yourself during your first tangle with the Mean Value Theorem.

  We'll see that there are several different kinds of examples; this basic one was constructed to make sense of *what a theorem says*. This one also answers in part a very common student complaint: "If I don't understand the theorem how can I draw a picture or do an example?" Answer: you can always draw and work with and try to apply the theorem to something as simpleminded as $f(x) = x^2$, $a = 2$, $b = 5$ or some other equally concrete example. The concreteness allows you to whittle away at the problem; you certainly could compute $f(5) = 25$ whether you understood anything else or not. Care in writing down what you can compute or do understand leaves you focused on what you don't, as does care in labeling pictures.

  Another part of the answer is that *any* picture or example is better than staring at nothing (the blank page). You may get to a better picture or example. You may realize that you don't understand a term in the theorem and won't make progress until you review a definition. A picture can take you quite a long way; work with a picture of the "generic" function is another route to understanding the Mean Value Theorem:

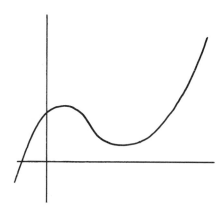

Start labeling: where is a? Where is b? $f(b)$? $f(a)$? Where does c have to be? What is $f'(c)$ or $(f(b) - f(a))/(b - a)$? How can they be included in the diagram? Once you touch the pencil to the paper you are already way ahead of the technique of staring.

1.3.1   Exercises

1.36: Theorem: If A, B, and C are sets, then

$$A \cap (B \cup C) = (A \cap B) \cup (A \cap C).$$

1.37: Theorem: If A, B, and C are sets, then

$$A - (B \cap C) = (A - B) \cup (A - C).$$

1.38: Theorem: If $\{\{x\}, \{x, y\}\} = \{\{u\}, \{u, v\}\}$ then $x = u$ and $y = v$. Remark that this theorem is the basis of one definition of <u>ordered pair</u> in terms of sets. In your example, consider the special case in which $x = y$. Also, would $x = v$ and $y = u$ work to make the sets equal?

1.39: Theorem: If R is a relation on a set S, and every element of S occurs as a member in at least one ordered pair in R, and if R is symmetric and transitive, then R is reflexive. Explore. Definitions you may need are in Exercise 1.21 and following.

1.40: Here is a definition.

Definition   *A <u>partition</u> of a set A is a collection of subsets $\{A_\alpha\}$ of A that are pairwise disjoint (that is, any pair of them has empty intersection) and so that A is the union of the A_α.*

For the set $\{1, 2, 3, 4, 5\}$ construct several different partitions. To start with, partition $\{1, 2, 3, 4, 5\}$ into sets A_1 and A_2. Note in passing that the name "partition" is entirely appropriate for the concept.

1.41: A partition $\{A_j\}$ of a set A is said to be <u>finer</u> than a partition $\{B_k\}$ of A if for every j there exists a k so that $A_j \subseteq B_k$. After you have explored this definition, consider the following theorem.

Theorem *Every set S has a finest partition.*

1.42: The following, the Intermediate Value Theorem, is from elementary calculus (although its proof is very far from being elementary calculus).

Theorem *Let f be a real-valued function defined and continuous on a closed interval $[a, b]$, and suppose $f(a) > 0$ and $f(b) < 0$. Then there exists a point c in the open interval (a, b) such that $f(c) = 0$.*

1.43: You may need to recall some definitions from Exercise 1.17; also, two paths from vertex v_1 to vertex v_2 are <u>independent</u> if they have no vertices in common except v_1 and v_2.

Theorem *If two vertices v_1 and v_2 are on a cycle then there exist at least two independent v_1–v_2 paths.*

1.4   Extended Examples

Let's move to another kind of example. Our square function example to go with the Mean Value Theorem is so simple that the theorem doesn't appear very powerful. You don't need a theorem to show that a c satisfying the conclusion exists since you can (and did) produce such a c. Can we construct an example in which the conclusion of the theorem seems reasonable but can't be verified (at least trivially) otherwise? Sure; we do need a function f so that the solution of the equation $f'(c) = r$ (where r is known) isn't easy (why?). If f is a cubic or quadratic or linear we would have f' quadratic or [6] which would be too easy. Some possibilities are higher-degree polynomials or nonpolynomial functions. Let's try with the sine.

Take f given by $f(x) = \sin x$, $a = 0$, and $b = \pi$. Then the theorem says there is a c in $(0, \pi)$ such that

$$f'(c) = \frac{f(\pi) - f(0)}{\pi - 0} = \frac{0 - 0}{\pi - 0} = 0.$$

That is, $\cos c = 0$. Is there a c in $(0, \pi)$ such that this happens? Rats! Yes, there is, and we can compute it explicitly since with $c = \pi/2$ we have

---

[6]Why am I doing this? You should be doing this.

$\cos c = 0$. This is a perfectly fair example but not one to show the power of the theorem by giving us something we can't get by hand.

There are two possibilities. We could abandon the sine function and try some other, or we could fiddle with a and b. Fiddling is easier and avoids starting over (don't discard the work already done without trying to get something from it). Further, our ability to solve $\cos c = 0$ was a matter of luck; solving $2c = r$ in our first example is easy for any r, but solving $\cos c = r$ was possible only because of an exceedingly lucky (or unlucky, depending on how you look at it) choice of a and b giving $r = 0$. What a and b might we try instead? Apply the theorem to a pair of your choice.

1.44:

One of many reasonable pairs is $a = 0$, $b = \pi/2$. (This one has the advantage of preserving as much of the work from our previous try as possible.) For this revised example, we hope that

$$f'(c) = \frac{f(\pi/2) - f(0)}{\pi/2 - 0} = \frac{1 - 0}{\pi/2 - 0} = \frac{2}{\pi}$$

holds for some c in $(0, \pi/2)$. Think of a good thing to do next to see if this looks plausible.

1.45:

Among the various good things to do[7] is to draw a picture.

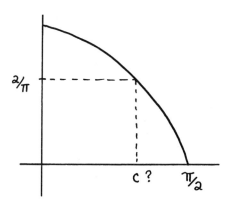

The conclusion of the theorem requires that

$$\cos c = \frac{2}{\pi}$$

---

[7]The only bad thing is to do nothing and keep reading.

for some c in $(0, \pi/2)$. It certainly does look as if such a c exists, and it is hard to see how to find it by hand. We've got the sort of example we've been seeking.

We might consider ourselves satisfied by intuition that a c exists, or we might try to see how to justify things further (without, of course, the use of the theorem, whose conclusions we are trying to test, not use; one justification is from the Intermediate Value Theorem, a version of which is in Exercise 1.42). But surely in this case the theorem gives us a "novel" conclusion.

What's the point of this kind of example? If the Mean Value Theorem is a good theorem (and it's in every calculus text) it must give you something you can't get easily otherwise. If not, why bother stating or proving it? Trivial theorems don't get a lot of press. Simple examples are the place to start understanding but don't usually show the full power of the result. Examples as above take you farther. It goes without saying that this sort of example is just as useful for definitions.

Aside: Natural History and Propaganda

For the definition of mammal the duck-billed platypus is an extended example. The platypus is native to Australia; has webbed feet, a furry body, and a bill resembling that of a duck; and the females nurse the young. So far so good (except, perhaps, for the bill). They also lay eggs. This came as a surprise to the naturalists who first discovered the animal; indeed, when the first specimen was sent to England, it was immediately pronounced a fraud, composed of pieces of various animals sewn together (and that doesn't even deal with the egg business). It may be surprising that they are included in the mammals, but given that they are (and they do nurse), why should one know this sort of example? At the very least, it shows that things we associate intuitively with mammals are not part of the technical definition.[8] The boundaries of what is allowed are farther out than we thought.

This sort of unconscious association of what is *common* with what is *required* happens in mathematics too, and some of the important examples in the history of mathematics served to address this very sort of intuitive expectation. Think, for example, of the relationship between continuity of a function at a point and differentiability of a function at a point. We know that "differentiable implies continuous" and we know that things don't go the other way. For example, the absolute value function is continuous at 0 but is not differentiable there. But, after all, the absolute value function *is* continuous everywhere else. If a function is continuous everywhere, doesn't it have to be differentiable "most of the time?" The construction by Weierstrass in 1874 of a function continuous at every point yet differentiable at

---

[8]One often sees an informal definition including "bears the young alive." Oh well.

no point showed that the condition "continuous" does not rule things out in the way we expect (an earlier example by Bolzano in 1834 didn't become known at the time). This shattering of the unexamined belief that "continuous functions are differentiable most places" was enlightening, signaled to the mathematicians of the time the success of some new methods, and launched some research into the better understanding of what continuity really does give you. A good deal of rich mathematics followed.

A similar example had to do with the notion of "curve." Memories of third-semester calculus might remind you that a continuous curve is the image in the x–y plane of a continuous function from some subset of the real numbers (say, an interval) into the set $\mathbf{R}^2$ of ordered pairs of real numbers. But anyway, you know one when you see one, right? The construction by Peano (among others) of a "space-filling curve" that, while the continuous image of the unit interval $[0,1]$, nonetheless filled the whole unit square $[0,1] \times [0,1]$ in the plane was an indication to the mathematicians of the time that they didn't know one when they saw one either. Again, this extreme example of a continuous curve sparked a lot of good mathematics. **End of Aside**

At this point we have two kinds of examples, the first being the simplest possible "getting started" sort. Let's agree to call these *basic* examples, and realize we have them for both theorems and definitions. They are what makes a theorem or definition readable. The second sort tries to push the theorem a bit to see whether the conclusion has some power, or to test the outer limits of the definition. Let's call these *extended* examples. The exercises below let you practice the construction of both kinds.

1.4.1 Exercises

1.46: Give some examples to test the outer limits of the definition of mammal. Some of them might show why things you might expect as part of the definition don't appear.

1.47: Recall (see Exercise 1.42) that the Intermediate Value Theorem from calculus is as follows:

Theorem 1.4.1 (Intermediate Value Theorem) *If f is a real-valued function continuous on $[a,b]$ with $f(a) > 0$ and $f(b) < 0$ there exists a number c in $[a,b]$ such that $f(c) = 0$.*

Explore this theorem again including some extended and non-examples.

1.48: Recall the definition of relation on a set from Exercise 1.21 and following. Using the set $S = \{1, 2, 3, 4\}$ of that exercise, can you find some extended examples of relations on S? What is the largest? What is the smallest?

1.49: If R is a relation from A to B (see Exercise 1.20), define R^{-1}, the <u>inverse of the relation</u>, to be the relation from B to A such that $(b, a) \in R^{-1}$ if and only if $(a, b) \in R$. Explore this definition thoroughly.

1.50: We will shortly become interested in relations R from A to B with the property that each element of A appears exactly once (once and only once) as the first coordinate of an ordered pair in R. Can you find a relation R from A to B with this property? One with it but such that R^{-1} is without it (that is, it isn't true that each element of B occurs exactly once as the first coordinate of an ordered pair in R^{-1})? So that R is without it but R^{-1} has it? Neither? Both?

1.51: Theorem: If A, B, and C are sets, then

$$A \cap (B \cup C) = (A \cap B) \cup (A \cap C).$$

This is a return to Exercise 1.36, but you should be able to produce some more examples now. For example, can you produce an example in which both sides are the empty set? One in which one term is the empty set?

1.52: Definition: A real-valued function f is <u>convex</u> on $[a, b]$ if for every t, $0 \leq t \leq 1$ and every x_1 and x_2 in $[a, b]$ one has $f(tx_1 - (1 - t)x_2) \leq tf(x_1) + (1 - t)f(x_2)$. Explore this definition with some examples. Warning: this one is fairly hard. If you don't use a picture, you are surely doomed. Can you explain the choice of the word "convex" for this notion?

1.53: A sequence $(x_n)_{n=1}^{\infty}$ is said to be <u>increasing</u> if for every n we have $x_{n+1} \geq x_n$.[9] Explore this definition with some examples. One of them ought to take full advantage of the "greater than *or equal to*" in the definition.

1.54: Suppose E is an equivalence relation on a set S (recall the definition from Exercise 1.25). For each x in S, define $E_x = \{y \in S : (x, y) \in E\}$. (That is, E_x is the set of all things equivalent under the relation E to x.) Proposition: Two sets E_x and E_y are either disjoint or identical. Explore with some examples.

1.5   Notational Interlude

Since we will be discussing functions a good deal in what follows, let us adopt the following notational conventions. When we discuss a function, say f, it is to be understood that we have in mind a certain <u>domain</u> [denoted *domain*(f)] and a certain <u>codomain</u> [denoted *codomain*(f)]. By

---

[9]Observe that this isn't really *increasing* but something like *nondecreasing*. This abuse of language is confusing unless you are used to it but convenient and hence common. Sorry about that.

codomain(f) we mean some fixed set such that $f(x) \in codomain(f)$ for each x in *domain(f)*. This is not what we shall call *range(f)* (the <u>range</u> of f), which is the set of all $f(x)$ such that $x \in domain(f)$. The codomain is therefore some fixed set containing the range. These definitions should be consistent with a careful definition of "function," which you will see soon if you haven't already. For the moment these definitions may be applied to your current understanding of function.

* * *

We hope against hope that you constructed various examples for these definitions. If not, you fell into the trap of "ordinary reading," which is a polite name for passive reading. It is easy to do; unlike most parts of this book, which give you a cue as to when to stop reading and start writing, this part didn't and most books don't. *You* have to make a habit of doing the needed things. Take some time to do them now.

1.55:

1.6   Examples Again: Standard Sources

Supposing you have been convinced that examples are useful, you may still be wondering where they come from: how do you build one when you need it? The best place to find examples is in the pool of examples you already constructed for something else, or perhaps a variant of one of them. But if your pool is small at the moment, and sometimes when you are just starting out on something new, you may have to build one from scratch. The goal of this section is to teach you where mathematicians look for examples. Good news: there are standard starting places.

Consider the following definition.

Definition   *Suppose f is a function and let B be a subset of codomain(f). Define the <u>pre-image</u> of B, denoted $f^{-1}(B)$, by*

$$(1.2) \qquad f^{-1}(B) = \{x : x \in domain(f) \text{ and } f(x) \in B\}.$$

Before exploring, realize that this definition yields an object. We are indeed associating a name with a condition, but the name is associated to the set containing all those things satisfying the condition. (It is a little as if we had defined not "function continuous at a point" but "the set of functions continuous at a point" instead.) Now go explore.

1.56:

After success with one B, this is a nice one to try with other sets and/or functions. But a good rule of thumb is to vary only one thing at a time.[10] In the search for extended examples, can you find, keeping your f, a set B so that $f^{-1}(B)$ is a pair of points? A single point? A closed interval? A pair of closed intervals? Can you give a B so that $f^{-1}(B)$ has no elements in it (is the <u>empty</u> or <u>null</u> set)? Can you describe all such sets B?

1.57:

1.6.1   Small Examples

We next give some standard places to find "good" examples and use the above definition as a testing ground. First, especially in dealing with functions and sets, what might be called *small* examples are useful. (For example, if you did the "relations" problems, we did it for you, and we have in fact used it repeatedly in the Hints in the Appendix.) Suppose we try to construct a function f with domain $domain(f) = \{a, b, c\}$ and $codomain(f) = \{1, 2, 3, 4\}$. How can one specify such a function? If you think back to your precalculus graphing, in which you probably plotted points, a function can be well specified by a table:

x	$f(x)$
a	2
b	?
c	?

(Here the marks "?" indicate that you need to choose some values.) Make and write down some choices for $f(b)$ and $f(c)$, and note that you have completely specified the function in so doing. You may now consider, for various B, the sets $f^{-1}(B)$. Almost all of the questions we asked above for your first example[11] may be asked here or have analogs. Ask and answer them!

1.58:

You can do more by modifying the above example after a bit of thought. Your choices for $f(b)$ and $f(c)$ were probably random; can you now make

---

[10]If you are playing with a new stereo system without the owner's manual, and you begin by pressing as many buttons as you can, all at once, you won't make much progress. What would you really do?

[11]Was it the square function? If not, you are either hopelessly confused or extremely independent. If you had no function at all, you are stunningly lazy.

different choices so as to produce an example with interesting properties (recall that we called this an *extended* example)? Remember that $codomain(f) = \{1, 2, 3, 4\}$. You might feel the impulse to take the definition of $f^{-1}(B)$ and bump it up against the definitions injective and surjective. Give in to that impulse.

1.59:

Can you make a conjecture — a guess — about how $f^{-1}(B)$ behaves if f is injective? Can you test it on examples of real-valued functions, including the one with which you started the process?

1.60:

The moral of the story is that a great deal of the behavior of this set $f^{-1}(B)$ can be displayed in very small, easily understood, and quite concrete examples.

Of course you drew pictures when using a real-valued function on this problem.[12] There are some good pictures for smaller examples as well:

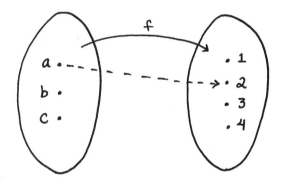

Drawing in B and $f^{-1}(B)$ may seem childish, but it really does help one's intuition. (In particular, it is impossible not to have to make a choice as to where to put B: domain or codomain? Of course only one of these is right, but without pictures you can fool around for a long time without ever confronting this choice.) Go get comfortable with this.

1.61: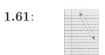

---

[12]Shame on you if you didn't! Make sure your picture clearly indicates that B is a subset of the y axis, since it is a subset of the codomain, and that $f^{-1}(B)$ is a subset of the x axis.

1.6.2 Exercises

1.62: Here is a definition.

Definition *Let X be a set and A a subset of X. The function χ_A with domain X and codomain $\{0, 1\}$ is defined by*

$$(1.3) \qquad \chi_A(x) = \begin{cases} 0, & x \notin A, \\ 1, & x \in A. \end{cases}$$

The function χ_A is called the <u>characteristic function</u> of A.

Note that this definition really defines a whole family of functions with common domain X, one function for each subset of X. You know what to do.

1.63: Here is another definition.

Definition *Let f be a function with domain X and Y some codomain. For B a subset of X, define the <u>image</u> of B under f, denoted $f(B)$, by*

$$f(B) = \{y : \exists x \in B \text{ such that } f(x) = y\}.$$

(Recall that "$\exists$" means "there exists.")

1.64: Here is a theorem.

Theorem *Suppose f is a function with domain X and Y a codomain. If f is both injective and surjective on Y, then for any element y of Y, $f^{-1}(\{y\})$ contains exactly one element.*

Explore this theorem thoroughly with examples.

1.65: We finally give a formal definition of function.

Definition *A <u>function</u> f from A to B is a relation from A to B such that if (a, b) and (a, c) are in f then $b = c$.*

Note that in a relation we do not assume that each a in A appears (at least once) as a first coordinate of a pair in the relation; with a function from A to B, it is customary to assume that A is the domain of the function, which means that indeed each a in A appears (at least once) as a first coordinate of a pair in the function. *Note that under this definition a function is a set of ordered pairs.* Go become extremely comfortable with this definition, beginning with small examples and moving to more familiar functions.

1.66: You probably saw many "definitions" of function something like "a function from A to B is a rule associating with each element of A a *unique* element of B." Of course this is not a definition, because "rule" is undefined.[13] The usual notation for functions ($f(a) = b$) is now simply defined to mean $(a, b) \in f$. With this notation in hand, can you see in the proper definition where the "*unique* element of B" idea is captured?

1.67: Since a function f is a (special kind of) relation, we can talk about f^{-1} for any function (see a definition in Exercise 1.49). This gives us a *relation* f^{-1}; there is no guarantee that it is, itself, a function. Find some examples where it is, and where it isn't. Find a relation R which is not a function but such that R^{-1} is a function.

1.68: For a graph G define the <u>distance</u> between vertices v_1 and v_2, denoted $d(v_1, v_2)$, to be the length (that is, number of edges) in the shortest v_1–v_2 path (if there are no v_1–v_2 paths, define the distance to be "infinity"). Define the <u>eccentricity</u> of a vertex v to be the maximum $\max_{u \in G} d(u, v)$. Explore. Can you find a graph in which every vertex has eccentricity one?

1.69: After you have finished the exercises above, go back and critique what you did. Did you construct basic and extended examples? Did you use both small examples and examples of real-valued functions? Do you have a full set of examples for the theorems? Did you make any conjectures?

1.6.3   Extreme Examples

A sort of generalization of the search for small examples is the search for *extreme* examples. Let us return to our testing ground, the definition of $f^{-1}(B)$. Among the B you should learn to choose are B as "*large*" as possible (for example, the whole codomain) and as "*small*" as possible (the empty set, or a set containing only one point). Or how about, for this example, B the set of all the points in the range, or all the points not in the range. In this class of extreme examples are, as well, very large or small sets for the domain or codomain [some f with *domain*(f) a set with only one element, for example]. Also, there are extreme choices of functions. One function that frequently shows unexpected behavior is the constant function.[14] This function is extreme in two senses: its range is as small as possible (just one point) and it is very far from being injective. The sine function is a real-valued function not constant but still very far from being injective and is also a nice one to try things on. The exercises below ask you to include this sort of examples in your packages for the set of exercises you just completed, as well as some others.

---

[13]But the teacher smiled, gave a lot of examples, and everybody sort of knew what was meant, right? And of course teachers never lie ... .

[14]This is a good one for shooting down overly simplistic conjectures, too.

1.6.4 Exercises: Take Two

1.70: Fill out the package for Exercise 1.62 of the last section.

1.71: Fill out the package for Exercise 1.63 of the last section.

1.72: Fill out the package for Exercise 1.64 of the last section.

1.73: Fill out the package for Exercises 1.65, 1.66, and 1.67 of the last section.

1.74: A group is a set with a "binary operation" [that is, a function taking two (hence, "binary") elements of the set and returning an element of the set; both "addition" and "multiplication" on the real numbers are examples] that satisfies certain algebraic properties (for example, associativity). We will call the binary operation $*$, so we write things like $a * b = c$. You may take on faith for this exercise that in any group there is a unique identity element (which we call e), so that for the binary operation we have $e * a = a * e = a$ for all a in the set. It is also true that in a group one has "left cancellation," so if $a * b = a * c$ one may deduce $b = c$, and also "right cancellation." Armed with these facts (which are by no means a full definition of group, but are enough to get by with here), determine the structure of the group with two elements. You may act on the (true) belief that it is enough to determine the operation table,

$*$	e	a
e	?	?
a	?	?

where the "one–two" position is $e * a$, for example.

Remark: this problem involves the construction of a different sort of "concrete example." If you try to find some familiar operation and some familiar things for e and a, you will almost certainly go wrong if you haven't had a group theory course. This is a sort of "concrete" *abstract* example, in that while e and a are abstract, we are at least trying to have the operation table completed in terms of them.

1.75: Determine, as in the previous example, the structure of the group(?s) with three elements.

1.76: We develop in this exercise another example which you will learn some day is an example of a group. The point is not to have you come up with an example completely independently, but, as in Exercise 1.74, to complete the details of one with a little help.
a) First we define some sets. Given some positive integer p greater than 1, define the set $\mathcal{O}_k$ for each k, $0 \le k \le p-1$, to be the set of all non-negative integers n such that n divided by p leaves remainder k. (Of course, we are talking about division among the integers, so there really are remainders.) Explore a little.

b) Next, we are going to define an "addition" operation "$\oplus$" on the sets $\mathcal{O}_k$ by defining $\mathcal{O}_i \oplus \mathcal{O}_j$ as follows. Pick n_i in $\mathcal{O}_i$ and n_j in $\mathcal{O}_j$, and define $\mathcal{O}_i \oplus \mathcal{O}_j$ to be the set $\mathcal{O}_k$ containing $n_i + n_j$. Explore; check that, although it looks as if you might get different answers for a certain $\mathcal{O}_i \oplus \mathcal{O}_j$ depending on which elements in them you picked to add, in fact you always get the same $\mathcal{O}_k$.

c) Define a "multiplication" called "$\otimes$" on these sets similarly. Can you find some p so that there are $\mathcal{O}_i$ and $\mathcal{O}_j$ so that $\mathcal{O}_i \otimes \mathcal{O}_j = \mathcal{O}_0$ with <u>neither</u> i nor j zero?

1.77: Produce a one-to-one correspondence between the sets $\{1, 2, 3, 4, \ldots\}$ and $\{2, 4, 6, 8, \ldots\}$. See Exercise 1.16 for the definition.

1.78: Recall the definition of <u>relation</u> on a set S from Exercise 1.21. Using again the set $S = \{1, 2, 3, 4\}$, can you find a "big" relation? A small relation? A small reflexive relation? A big and small equivalence relation?

1.79: Continuing the above exercise, can you find a big relation from A to B? A big function from A to B? Small ones of each?

1.80: What is the graph on n vertices with the largest number of edges? Smallest? Also, recall from Exercise 1.68 the definitions of distance and eccentricity for a graph. Can you find a graph on 4 vertices with a pair of vertices so that the distance between them is not infinity but is still as big as possible? What about eccentricity for a graph on 5 vertices; how large can that be? How small?

1.7 Non-examples for Definitions

There's another good class of examples (which we have run into a little before). As part of a theorem or definition package one needs some "non-examples" (for lack of a better name). These are easiest to understand for definitions: a lizard is a non-example for the definition of mammal. That seems simple enough, but the shark[15] is also a non-example, and it is worth seeing why either alone is less useful than the pair. The definition of mammal has a number of conditions to be met, and a lizard is a simple non-example because it fails just about every one of them. The shark is an extreme non-example because it fulfills many of the conditions (but not all of them). It shows why the definition has to include so many conditions: if the definition were to include only some conditions, the shark would not be excluded.

---

[15]Sharks are warm blooded, but they do not nurse their young. Also, they don't have vertebrae, but simply a spinal chord. The difference between a shark and a dolphin (which is a mammal) is nonetheless not immediately apparent.

Faced with a mathematical definition one may play the same game. It is certainly worthwhile to have something not satisfying the definition. But further, if the definition has several conditions, drop one of them while keeping the others and try to cook up an object to fit. If it is impossible to come up with such an object the condition in question is redundant (taken with the others) and may be omitted from the checklist without changing things. If not, you have found a non-example whose bizarre behavior the definition desires to exclude, and you may see why it is nicer not to allow such monsters. You may then repeat the game with a different condition dropped to find a new unusual animal.

You may object to the description of the above as a "game," since games are supposed to be fun. Leaving out the fact that for many mathematicians this *is* fun, what is the point if we have to be serious? The various conditions of a definition need to be understood separately and together, and the effort to construct examples forces you to do that.

Let's (meaning *let us*, meaning *let you*) work through an example with one of the standard calculus definitions.

Definition *A real-valued function f is said to be <u>continuous</u> at the point a if*

1. *f is defined at a,*

2. $\lim_{x \to a} f(x)$ *exists, and*

3. $\lim_{x \to a} f(x) = f(a)$.

Well, begin to construct some non-examples for this definition. This is a good problem to attack with pictures first. Start by dropping condition (1).

1.81:

Welcome back. Now you may start on what happens when condition (1) is retained but condition (2) is dropped [again we lose (3) as well].

1.82:

If you are distressed that we are merely drawing pictures, a useful device for constructing functions avoiding condition (2) at some point is the sort of function defined piecewise as in the example below:

$$f(x) = \begin{cases} x^2, & x \geq 0, \\ 4x + 5, & x < 0. \end{cases}$$

If you *have* to write down a single formula, try playing with the function whose formula is $x/|x|$.[16]

1.83:

Finally, what happens if condition (3) alone is deleted?

1.84:

Your collection of examples should have shown you that the functions allowed by definitions with parts missing are substantially worse than those allowed by the full definition. Further, and very useful for things to come, you have the full package for the definition "continuous at a point." The collection of examples you've found will be reused. Trust me.

The exercises that follow give you a chance to practice this sort of non-example for definitions.

1.7.1 Exercises

1.85: Here is a definition.

Definition *A sequence $(x_n)_{n=1}^{\infty}$ is <u>bounded</u> <u>above</u> if there exists some number M such that $M \geq x_n$ for all n.*

Explore this definition. Of particular interest are sequences both increasing and bounded above. (Recall that increasing was defined in Exercise 1.53 of Section 1.4.1.) Explore this pair of definitions with basic and nonexamples.

1.86: Here is a definition basic to much of abstract algebra.

Definition *A <u>binary</u> <u>operation</u> on a set S is a function from $S \times S$ to S.*

While this definition is precise, it is not what we usually think of as an operation. Begin by seeing how the "operation" of adding two numbers can be thought of this way. Produce some other examples from your previous mathematical history. Can you produce some non-examples? Finally, if the set of objects is the collection of real-valued functions, can you produce some binary operations on the functions?

---

[16]But the insistence upon viewing a function as something coming from a formula is limiting in ways that may not have been apparent yet. Just as all relations are not as easily describable as "<" is, not all functions come with a neat formula. This is especially true when you want to construct "bad" functions to use as non-examples.

1.87: Since many binary operations are like the standard example of addition, we tend to write a binary operation in that sort of notation: instead of calling the binary operation B and saying $B(x, y) = z$, we tend to call it something like $*$ and write $x * y = z$. With this convention, we may make some definitions.

Definition 1.7.1 *A binary operation $*$ on a set S is <u>associative</u> if, for every x, y, and z in S, $(x * y) * z = x * (y * z)$.*

Definition 1.7.2 *A binary operation $*$ on a set S is <u>commutative</u> if, for every x and y in S, $x * y = y * x$.*

Definition 1.7.3 *Suppose $*$ is a binary operation on a set S. We say e is an <u>identity</u> <u>element</u> for $*$ if, for every x in S, $x * e = e * x = x$.*

Definition 1.7.4 *Suppose $*$ is a binary operation on a set S with an identity element e. We say an element x has an <u>inverse</u> with respect to $*$ and e if there exists an element y of S such that $x * y = y * x = e$.*

Remark that under the sorts of binary operations ordinarily studied, if there is an identity element for $*$ there is only one, and so in the definition of inverse element one can delete the reference to "inverse with respect to e," since there is only one identity element you could possibly have an inverse with respect to.

Explore these definitions fully, using every binary operation you can think of. It is also worth pausing a moment to see why these names are reasonable for the named concepts.

1.88: Recall the definition of relation on a set S, and of reflexive, symmetric, and transitive relations from Exercises 1.21 and following. Fill out your packages for these definitions with some non-examples. There are also more definitions associated with relations: a relation R on a set S is <u>irreflexive</u> if for each $a \in S$, $(a, a) \notin R$. A relation is <u>antisymmetric</u> if, for each a and b in S, $(a, b) \in R$ implies $(b, a) \notin R$. Produce complete packages of examples and non-examples for these new definitions. A test of how full your packages are is to see whether you can produce a relation that is neither reflexive nor irreflexive, and one that is neither symmetric nor antisymmetric. Again, convince yourself as well that the mathematicians who named these concepts used helpful names.

1.89: Here's another definition.

Definition *Assume that $(a_n)_{n=1}^{\infty}$ is a sequence of positive numbers. Then the series $\sum_{n=1}^{\infty} (-1)^n a_n$ is <u>alternating</u>.*

By the way, why is it called "alternating"?

1.90: A graph is said to be <u>connected</u> if every pair of vertices has a path between them.

1.91: A <u>perfect</u> <u>pairing</u> of a graph G is a collection P of ordered pairs of vertices such that each vertex occurs in exactly one of the pairs and such that for any pair (v_1, v_2) in P the edge v_1-v_2 is actually in the graph G.

1.92: The <u>degree</u> of a vertex of a graph is the number of edges of which it is an end point. Explore. A (finite) sequence of integers is called <u>graphic</u> if it is a list of the degrees of all the vertices of some graph. Explore. In particular, can you find a sequence which is *not* graphic?

1.8   Non-examples for Theorems

There is another useful class of non-examples, this time for theorems. Well, the goal would appear to be to find something not fitting the theorem in question. For example, a cow doesn't fit the Mean Value Theorem (as a former student, asked for a non-example, pointed out to me). This true but not useful remark should make it clear that we need to specify a little better what we want.

Recall that a theorem has two main sections (well, all right, most theorems). One consists of the condition of the hypotheses (assumptions), both explicit and implicit.[17] The other piece is the (condition which is the) conclusion: the thing that an object satisfying (all of) the hypothesis is guaranteed to do in addition. It would seem that the following combinations of success and failure on the two pieces might be interesting:

Hypothesis	Conclusion
YES	YES
YES	NO
NO	YES
NO	NO

Any sort of active reading will convince you one of these is a dummy and can't occur. Which one?

1.93:

The others are at least logically possible, and the basic and extended examples discussed earlier address the case in which both hypothesis and conclusion are satisfied.

Think back to the Mean Value Theorem to see why examples for the other possible cases (which are therefore non-examples) might be useful.

---

[17]For example, one might state the MVT without clearly noting that some object f is a function, but the assumption is still there.

The basic and extended examples for the MVT we found in the last section indicate something about what the theorem says and why it might be worth saying. But *why* the theorem is true (that is, are the hypotheses necessary, and how do they work to produce the conclusion?) is not so clear. One answer to these questions is, of course, the proof, but some well-chosen examples and pictures can help too. (Also, sometimes the proof isn't very helpful. Proofs by contradiction, for example, tend to show why something *isn't false* rather than why it *is true*.)

Suppose we can find a non-example in which the hypothesis is not satisfied and the conclusion isn't satisfied either (the NO–NO case above). The example will probably show us something about how the hypothesis is related to the conclusion. Indeed, if the hypothesis has several conditions we can do what we did with definitions (violate them one at a time) and see if we may still evade the conclusion. Such an example might fall into some class of "non-examples for weakenings."[18] It would be a counterexample for a proposed theorem with weaker hypotheses. The collection of the basic, extended, and non-examples we get out of this is the package for the theorem.

Let's use the Mean Value Theorem as the practice field. The dissection of the formal language of the MVT into a useful form will be discussed thoroughly in a chapter to come. Suppose for the moment this has been done to yield the following (nonstandard but useful) version:

Theorem 1.8.1 (Mean Value Theorem II) *Suppose f is a function on $[a, b]$ satisfying*

  1. f is differentiable on (a, b), and

  2. f is continuous at a and at b.

Then there exists a point c in (a, b) such that

$$f'(c) = \frac{f(b) - f(a)}{b - a}.$$

We may start by seeing whether, for example, if we keep assumption (2) but dispense with (1) the theorem still holds [that is, assumption (2) alone forces the conclusion]. You might guess that it need not (unless you think millions of calculus books are peddling inferior theorems). If this guess is correct, somewhere out there should be a function f and an a and b such that f is continuous at a and at b yet there is *no* c in (a, b) such that

$$f'(c) = \frac{f(b) - f(a)}{b - a}.$$

To find such an f we had better confine our attention to those f not differentiable on (a, b). Why?

---

[18]Suggestions for better names eagerly accepted. Modest reward offered.

1.94:

This observation and the fact that "f not differentiable on (a, b)" means "there is at least one point of (a, b) at which f fails to be differentiable" tell us to look for an f such that

1. f is not differentiable at (at least) one point of (a, b), and

2. f is continuous at a and at b, and

3. there is no point c in (a, b) such that

$$f'(c) = \frac{f(b) - f(a)}{b - a}.$$

(The above analysis strays again into the upcoming chapter on formal language. Accept the above as the goal, but do try to see why it would show that hypothesis (2) of the theorem is not enough alone to guarantee the conclusion. Even the lack of one part of the hypothesis would make the conclusion not guaranteed by what was left.)

Where is such an f to come from? It must come from your package for the definition <u>differentiable</u>, in particular the non-examples section. Here is a fine opportunity to practice that sort of construction.

Miniexercise

Produce a variety of functions on various intervals that fail to be differentiable at one or more points. Pictures are quite acceptable, but do include at least one explicit function, and remember that the broader your collection of examples, the better. Go to it.[19]

1.95:

Welcome back. Recall that what you found are, for our purposes, functions that might satisfy (1) in the list above of three things our non-example function should do. Your diligence is very likely to be rewarded. **LOOK** at your pictures; does any one of them appear to satisfy (2) and (3) of that list?

---

[19] There are two observations here. First, we have broken away from our consideration of the MVT to do a subtask. Real mathematics, except for those very few mathematicians with total recall, is like this. Second, you really have to do this. This book, and the method it tries to present, is not for those who want to read without stopping.

1.96:

Probably, you will be convinced on intuitive grounds that no such c exists. Good; you have just constructed a non-example to show that hypothesis (1) may not be done away with if the conclusion of the MVT is to be guaranteed.

[If none of your examples satisfies (3), in the sense that there is (annoyingly) a point c satisfying

$$f'(c) = \frac{f(b) - f(a)}{b - a},$$

try moving a or b so that the c for your example is excluded.[20] Now is there a c for the new pair a and b? If there still is, try choosing an a and b very close to the point at which the function fails to be differentiable; there ought not to be a c in this case. If all else fails, try the absolute value function with $a = -1$ and a b of your choice.]

Don't stop; there's more to get from what you have done. First, you should have several examples of nondifferentiable functions. Does each of them serve as an example of a function satisfying (3) (and thus evading the conclusion of the MVT)? If not, can you modify an example somewhat so that it does? If you have an example that does not satisfy the hypothesis of the MVT but for which a point c does exist, hold on to it and we'll use it later.

Second, observe that you can construct an example of an f, a, and b satisfying (1) with f failing to be differentiable at only *one* point of (a, b). Indeed, one of your examples probably does this already, since it is hard to draw functions that are nondifferentiable at lots of points, and one tends to draw functions with a single "bad" point. Your example shows that the following non-theorem isn't true:

Non-theorem 1.8.2 (Almost MVT flop) *Suppose f is a function defined on $[a, b]$. If*

1. f is differentiable at all but one point of (a, b), and

2. f is continuous at a and b,

---

[20]We have used this device before. Pólya, in *How to Solve It* [5, page 208], has a "traditional mathematics professor" say "A method is a device which you use twice."

then there exists a point c in (a, b) such that

$$f'(c) = \frac{f(b) - f(a)}{b - a}.$$

This is very close to the Mean Value Theorem, yet not true. Your non-example shows that the full strength of hypothesis (1) seems to be vital for the success of the theorem.

Exercise

1.97: Find a function f defined on $[a, b]$ that is differentiable at all but one point of (a, b), and that is continuous at *all* points of $[a, b]$, but for which the conclusion of the MVT fails. (The force of this example is that while it is true that one could fail to be differentiable at a point by failing to be continuous at that point, such a "dramatic" failure of differentiability is not needed to shoot down the theorem.)

Where are we? For the Mean Value Theorem, hypothesis (1) seems really to be needed. What about hypothesis (2)? Go to it.

1.98:

Try to construct a non-example for which f is at least defined at the points a and b, and perhaps even continuous at one of them. (Quick question: could it be continuous at both?) Again, we seek to show that *any* failure of the continuity condition, not just a violent one, is enough to destroy the theorem.

Here are some exercises with which to practice the various kinds of examples.

1.8.1 Exercises

1.99: Try the fuller arsenal of things to do on the Intermediate Value Theorem (1.4.1).

1.100: Theorem: If A and B are sets with $A \subseteq B$ and $B \subseteq A$ then $A = B$. Explore fully, please, not just with non-examples.

1.101: The following is a standard theorem in first-term calculus.

Theorem 1.8.3 (Maximum Theorem) *Let f be a function continuous on a closed interval $[a, b]$. Then f attains a maximum value; that is, there is an x_0 in $[a, b]$ such that $f(x_0) \geq f(x)$ for all x in $[a, b]$.*

Explore this theorem. Note that the condition in the hypothesis is really a bundle of two conditions, one on the function and one on the set (this latter is easy to miss).

1.102: Here's a theorem from second-term calculus.

Theorem *Assume* $(x_n)_{n=1}^{\infty}$ *is an increasing sequence bounded above. Then* $(x_n)_{n=1}^{\infty}$ *has a limit.*

Explore this theorem. A definition you need is in Exercise 1.85.

1.103: Suppose f and g are functions for which the composition $g \circ f$ makes sense. Theorem: If $g \circ f$ is injective, then f is injective.

1.104: Theorem: Suppose R is a relation between A and B such that

1. R is a function,

2. R is injective, and

3. R is surjective on B.

Then R^{-1} is a function, injective, and surjective on A. Here's a hint: It will help a lot to think of things in terms of the definition involving ordered pairs. Also, it is probably unwise to tackle this one at all unless you have done previous exercises on relation, injective, and surjective.

1.105: Proposition: A walk with no repeated vertex (on a graph) contains no repeated edge.

1.8.2   More to Do

There is more to be done with non-examples for theorems. Recall that there is still the "NO–YES" part of the table to consider, in which we seek an object failing to satisfy the hypothesis of the theorem but satisfying the conclusion anyway. Recall also that during our discussion of the Mean Value Theorem you may have found a function f failing to be differentiable on (a, b), continuous at a and b, for which there did exist a point c satisfying the conclusion of the theorem. (This wasn't what we were looking for at the time, but we asked you to hold onto it.) If you didn't find one then, find one now.

1.106:

What is the point of these non-examples? Remember that a theorem is a guarantee that anything satisfying the (condition of the) hypothesis must also satisfy the (condition of the) conclusion. The theorem makes no statement whatsoever about objects not satisfying the hypothesis; in particular, it does not *prohibit* them from satisfying the conclusion. But it is interesting to see if in fact the conclusion may be obtained *without* satisfying the hypothesis. If that is not possible, there really is a companion theorem whose form is "conclusion of old theorem guarantees hypothesis of old theorem." That's worth knowing.[21] If it is possible to achieve the conclusion without satisfying the hypothesis, that's worth knowing too.

Try this on the following.

1.8.3 Exercises

1.107: Consider the Intermediate Value Theorem (Theorem 1.4.1).

1.108: Consider the Maximum Theorem (Theorem 1.8.3).

1.109: Consider the following theorem on alternating series (which uses a definition from Exercise 1.89):

Theorem *Suppose $a_n > 0$ for each n, so $\sum_{n=1}^{\infty} (-1)^n a_n$ is an alternating series. If also $a_n > a_{n+1}$ for all n, then the series converges.*

1.110: Theorem: If A and B are sets with $A \subseteq B$ and $B \subseteq A$ then $A = B$. Finish the exploration started in Exercise 1.100.

1.111: Continue the exploration begun in Exercise 1.102.

1.112: Continue the exploration begun in Exercise 1.103.

1.113: Continue the exploration begun in Exercise 1.105.

1.9 Summary and More Propaganda

The message of this chapter is that there is a way to help you learn mathematics independently and actively, namely to search for and construct examples. There are several classes of examples: basic examples, to get you started; extended examples, to help find the boundaries of the theorem or definition; non-examples, to help separate conditions and find relationships among them. There are standard places to look for examples, among them small and extreme ones. Examples, once found, have a way of turning up again in new circumstances, as do the techniques by which you construct and modify examples.

---

[21]Mathematics has been done too long for someone not to notice this little game. So there are theorems, using language like "if and only if" or "the following are equivalent," that capture this fact. In this case, hypothesis and conclusion are essentially bound together, in that where you find one, you also always find the other.

You might hope, but not really believe, this makes everything easy. But it is possible (if uncommon) to be so lost that you can't even start. More often, you may owe a debt to the past in that your past mathematics, learned less actively, is not well understood or accessible. (One of the reasons the approach of this text, or any active approach, is hard to start is that it tends to shine a merciless spotlight on your habits of the past and their results for your comprehension.) But the alternatives to this method work even less well. The "technique of staring" is a favorite unproductive method and so is the "hundreds of rereadings plus osmosis" approach, both of which give a surface understanding at best. Professional mathematicians talk about "getting your hands dirty," and it is true if uncomfortable that you have to.

Let us try to counter an objection. If you have been diligent in doing the tasks indicated in the text, a little voice is probably already saying, "Holy cats (or something worse)! If I have to do this for every definition or theorem, each *page* of mathematics with three definitions and a theorem is a six-hour assignment — no way!" There are two answers: first, the understanding you get this way is necessary. Think back to your calculus text, which was so long precisely because it tried to do all of this for you, and on paper at that.[22] How can you do new, as opposed to routine drill, problems if your understanding isn't this good? If you build examples regularly, you will find that you get good at it and it isn't too much more time consuming than what you used to do.[23] It may be hard right now to see this as enough compensation.

The second answer to the student complaint is that you *don't* have to do this whole process *every* time. We've been behaving as if your job were to re-invent all of mathematics from the ground up, righteously ignoring even the most nifty examples constructed by others. You don't. You need to use the technique whenever you are confused. You need to use the technique when you think a theorem is "obvious" or don't see how a definition could be any other way. It should be used sufficiently often so you are not in a fog six weeks from now (for example, if you are six weeks into real analysis and you think "all functions are continuous, so what's all the fuss?" you are completely lost). It should also be used enough so that there *is* some *invention* of mathematics in what you do, as opposed to merely receiving the word from the stone tablets owned by the professor. Invention and

---

[22]It is not uncommon to have seven or eight or more examples presented in a single section of one of the standard calculus texts. This is a feeble, and usually unsuccessful, attempt to get the effect of *your* construction of one or two good examples.

[23]The time is, however, spent differently. Instead of racing through the section, glancing at the book's examples, and learning all the material as you struggle to imitate the examples well enough to do the problems, you now learn the material as you go and prepare yourself for the problems before you get there. If teachers were given a guarantee that you did this, they could assign far fewer problems.

newness are fun and productive, and you now have the beginnings of a way to inject them into the mathematics you do.

1.9.1 Exercises

1.114: Recall the following theorem from first-term calculus.

Theorem (Squeeze or Sandwich Theorem) *Suppose f, g, and h are functions defined in some open interval I containing the point a. Suppose also that*

1. *$\lim_{x \to a} f(x) = \lim_{x \to a} h(x)$, and*

2. *$f(x) \le g(x) \le h(x)$ for all x in I.*

Then $\lim_{x \to a} g(x) = \lim_{x \to a} f(x)$.

This is not an easy one, but it is one for which the technique can yield a rich crop of examples. One thing to think about near the end is, need the functions actually be defined at the point a?

1.115: Here is a fairly straightforward definition:

Definition *A _partition_ of a set S is a collection of pairwise disjoint subsets A_α of S such that S is the union of the A_α.*

Explore. Now here is an important theorem.

Theorem *Suppose E is an equivalence relation on a set S. For any x in S, denote by E_x the set of all y in S equivalent under E to x. Then the collection of all E_x is a partition of S.*

Explore. This second half rests on Exercises 1.21 and following, 1.24, 1.25, 1.48, 1.54, and 1.88.

1.10   What Next?

In the next chapters we will turn to something to accompany the use of examples to help you with your mathematics (namely, informal and formal language). This is but one choice of many, though; there are further things to round out your understanding of definitions and theorems we could discuss instead.[24] Return to the definition of mammal to see what some of these might be. Surrounding the concept of mammal we identified a cloud of examples (of various kinds) and non-examples. But along with these you have the concept of "reptile". This is another definition at about

---

[24]See the Theoretical Apologia appendix for some language useful in talking about this process of rounding out a concept.

the same level as "mammal"; it isn't just a non-example like a lizard that happens to be a reptile, it is a parallel concept. It has its own cloud of examples, of course, but you also have the relationship between the class "mammal" and the class "reptile" (namely, that they are disjoint, since there is nothing that is both a mammal and a reptile). For a mathematical example, along with the definition of function continuous at a point (and its cloud of examples), is a parallel definition of function having a derivative at a point. One could write a whole chapter on the importance of making these connections along with fleshing out a concept with examples.

Or, we could note that along with your definition of mammal you have the idea that this concept fits into a larger picture (say, the division of living things into plants and animals). "Mammal" is one of a number of subdivisions of a larger scheme (and now, of course, it is clear where "reptile" goes). You know that along with this upward inclusion in structure is a downward branching into various subdivisions of mammals (carnivore, herbivore, omnivore, for example). We leave you to write the chapters on how to look for these connections and how to develop the techniques to do so, but realize that your experience in adding examples to the picture is good training for those tasks too. Examples are just the beginning . . . .

2
Informal Language and Proof

The previous chapter was all about active reading; while we used examples from mathematics, the techniques in that chapter are applicable almost anywhere. But reading mathematics has its own special problem: reading proofs. And after a while, you have to write your own. Reading and writing proofs rests on language, and that language is supposed to give you clues to what is going on both locally and in the proof as a whole. We start with ordinary English, which mostly gives *local* information; we'll later turn to proof structure, which is the frame for the big picture.

2.1   Ordinary Language Clues

We begin this section nontraditionally, with a set of exercises. Your short-term goal is to reorder the sentences in each of the following to assemble a correct proof. In most of the exercises, you may not know anything about the particular mathematics involved and so will have to rely on the language in which the proofs are written to provide clues to the proper order. The long-term goal is to see, along the way, if you can extract some rules about how the language can provide clues to what's going on in the proof.

2.1.1   Exercises

2.1: Prove that any triangle with two congruent sides has two congruent angles. (We use $\cong$ to denote congruent.)

1. Construct the segment AD.

2. Hence using standard facts about congruent triangles, we have angle ABC congruent to angle ACB, and we are done.

3. Let D be the midpoint of side BC.

4. Let A, B, and C denote the three vertices of the triangle.

5. Therefore, using Side–Side–Side and $AD \cong AD$, $BD \cong DC$, and $AB \cong AC$, we have triangle ABD congruent to triangle ACD.

6. Observe that $BD \cong DC$ by the definition of midpoint.

7. Suppose AB and AC are congruent.

2.2: Prove that the set of polynomials whose value is zero at $x = 1$ is a vector space.

1. By definition, $(\lambda p)(1) = \lambda p(1)$.

2. Therefore we are done from the theorem.

3. To check closure under addition, let p and q be two polynomials such that $p(1) = 0$ and $q(1) = 0$.

4. From the last two steps, and $\lambda \cdot 0 = 0$, we have $(\lambda p)(1) = 0$ as required for closure under scalar multiplication.

5. For closure under scalar multiplication, let p be a polynomial such that $p(1) = 0$ and let λ be any scalar.

6. We will use the theorem stating that since the collection of all polynomials is a vector space, we need only check for this (nonempty) subcollection that it is closed under addition and scalar multiplication.

7. Hence from the two previous steps, $(p + q)(1) = 0 + 0 = 0$, and we have closure under addition.

8. Note $(p + q)(1) = p(1) + q(1)$ by definition.

2.3: Prove that in a group any left inverse and any right inverse of a particular element are equal.

1. By the definitions, we have $ba = e$ and $ac = e$.

2. Using substitution with these three equations, and the definition of identity, we have $b = be = b(ac) = (ba)c = ec = c$, as desired.

3. Let e be the identity of the group, and let a be any element in the group.

4. Also, using associativity, $b(ac) = (ba)c$.

5. Suppose b is a left inverse of a and c is a right inverse of a.

2.4: Prove that for any n,

$$\sum_{k=1}^{n} k^3 = \frac{n(n+1)(2n+1)}{6}.$$

1. But $\sum_{k=1}^{n+1} k^3 = \sum_{k=1}^{n} k^3 + (n+1)^3$ by definition of sum.

2. We shall use induction on n.

3. To prove the "induction step," we assume for some n that

$$\sum_{k=1}^{n} k^3 = \frac{n(n+1)(2n+1)}{6}$$

and must prove that

$$\sum_{k=1}^{n+1} k^3 = \frac{(n+1)((n+1)+1)(2(n+1)+1)}{6}.$$

4. Therefore, from transitivity of equality and these three equations, we have

$$\sum_{k=1}^{n+1} k^3 = \frac{(n+1)((n+1)+1)(2(n+1)+1)}{6}$$

as required to complete the induction step.

5. Using substitution and our assumption, we have

$$\sum_{k=1}^{n+1} k^3 = \frac{n(n+1)(2n+1)}{6} + (n+1)^3.$$

6. To prove the case for $n = 1$, we must simply check that

$$\sum_{k=1}^{1} k^3 = \frac{1(1+1)(2 \cdot 1 + 1)}{6},$$

which is a trivial computation.

7. A computation shows

$$\frac{n(n+1)(2n+1)}{6} + (n+1)^3 = \frac{(n+1)((n+1)+1)(2(n+1)+1)}{6}.$$

2.5: Consider the sequence $(s_n)_{n=0}^{\infty}$ where s_n is defined by

$$s_n = \sum_{k=0}^{n} \frac{1}{k!}, \qquad n = 0, 1, \ldots.$$

Prove that this sequence converges.

1. Note also that, for each n, we have by an easy computation that $s_n \leq t_n$.

2. Define $(t_n)_{n=0}^{\infty}$ by

$$t_0 \;=\; 1,$$
$$t_n \;=\; 1 + \sum_{k=0}^{n-1} \frac{1}{2^k}, \qquad n = 1, 2, \ldots.$$

3. To show $(s_n)_{n=0}^{\infty}$ is bounded above, we employ an auxiliary sequence $(t_n)_{n=0}^{\infty}$.

4. Therefore, by the theorem mentioned above, we have $(s_n)_{n=0}^{\infty}$ convergent.

5. Observe that $(t_n)_{n=0}^{\infty}$ is convergent, since it is simply the sum of the constant sequence whose value is one and a familiar sequence that is the sequence of partial sums of a geometric series.

6. To show that $(s_n)_{n=0}^{\infty}$ is monotone increasing, observe that for any $n \geq 0$ we have

$$s_{n+1} = s_n + \frac{1}{(n+1)!} \geq s_n.$$

7. Since $(t_n)_{n=0}^{\infty}$ is convergent, it is bounded above, say by M, so $t_n \leq M$ for all n.

8. We shall use the theorem stating that a monotone increasing sequence which is bounded above converges.

9. From the two preceding inequalities, we have $s_n \leq M$ for each n, so $(s_n)_{n=0}^{\infty}$ is bounded above.

2.1.2 *Rules of Thumb*

Before looking at *our* rules of thumb for the use of language in proofs, see what *you* can extract from your experience with the above problems. Some questions to ask: how are "hence," "therefore," "thus," and "so" used? What about "also," "note," and "observe"? What is a "user-friendly" rule for the use of variables? How does one signal how the overall course of a proof is going to go?

2.6:

Perhaps your efforts produced something like the following:

Rules of Thumb for Proofs

1. **Tell the reader, right up front, what the general course of the proof will be.**[1] It is important to mark clearly this large-scale or global proof structure. This holds not only for the whole proof, but for subproofs in the body of the main proof, if there are any. The goal of this is to give the reader a set of (correct!) expectations for what is to follow.

2. **Introduce variables (names for objects) <u>before</u> you use them.** "Let," "denote," and "set" are some of the words marking this action.

3. Other "little words" are also **signposts** to the *small-scale* or *local* course of the proof.

a. **"Hence," "therefore," "thus," "so,"** and sometimes "then" have three pointers: a **forward, explicit** one to a *statement following*; a **backward, implicit or explicit** one to a *statement preceding* (or statements preceding); and an **implicit or explicit** one to the *reason* the following statement may be deduced from the preceding statement. The most complete use of these words makes all three pointers explicit. By convention, if no preceding statement(s) are indicated, the reader assumes the immediately preceding statement. The explicit mention of the reason for the deduction may be left out if it is "obvious" (whatever that means). So a pictorial template might be

$$\text{(reason)}$$
$$\text{(preceding statement)} \longleftarrow \overbrace{\text{THUS}} \longrightarrow \text{following statement}$$

b. **"Note," "observe," and "recall"** signal the collection of information for later use (recall is used for collecting some external fact from the general information base relevant to the proof, or for collecting some fact from earlier in the proof). They need not point at all, but may point backward to the source of the information.

c. **"Also" and "further"** signal the collection of (more) information for later use and carry the strong implication that several pieces of information will be used together soon.[2] ["Further" can also, unfortunately, be used as a signal for a follow-up deduction: Hence (something). Further, (something else).] They need not point at all, but may point backward, sometimes to a common source of several facts needed in what follows.

---

[1]Indeed, sometimes it's fine to mention it more than once. The "rule of three" — tell them what you're going to do, do it, and tell them what you did — is not a bad one.

[2]It's actually almost dramatic. If you read "Note blah. Also, blip. Further, blap" don't you expect a resounding "Hence **Dah-dah-dah**"?

4. A proof, even well marked along the way, is much easier to read if its global or large-scale structure is one of a few **standard forms** (one example is "proof by induction"). From the statement of how the general course of the proof will go (see 1), the reader gets a helpful double set of expectations as to what is coming, some from what you say about this proof in specific, and some from past proofs of this kind.

Before we talk about these rules, some exercises are appropriate.

2.1.3 Exercises

2.7: Go back to the previous set of exercises (2.1.1) and try again on those you weren't able to finish before.

2.8: From that same set of exercises, find several examples of the use of each of the rules mentioned above.

2.9: In Exercise 2.1 and Exercise 2.3 of that collection, one of the rules was violated. Which one? Could you untangle the proofs anyway? Why, or why one and not the other?

2.1.4 Comments on the Rules

Perhaps the best way to start the comments is with a question. What is the point of these rules?

2.10:

We hope you said something like this: the goal of these rules is to make the reading of the proof as easy as possible. You can't make the mathematics easier by good writing, but you can keep the presentation from getting in the way.[3] That goal sounds so obvious as to seem trivial, but it is important. The reader should be given correct expectations about what is to come, clear pointers to where we are coming from and where we are going, clear markers as to the task at hand and how it fits into the whole, and so on. The hope is to make the proof's presentation as simple as possible so that the reader's attention may be reserved for following the argument. A well written proof may still evoke a response like "I don't see how this step follows from this one." It should *not* evoke a response like "I don't see where this step comes from."

One other thing these rules do, which is important psychologically if not mathematically, is *reassure* the reader that all is going well. Half of reading

---

[3]As easy as possible may still not be easy, but a badly written proof creates unnecessary obstacles.

any argument is the belief that you can, and the reader of an induction proof who spots a friendly "For the induction step . . ." or another reader who sees "n" behaving as an integer is comforted that things are going well and encouraged to keep going. A reader who doesn't know where a step comes from (let alone whether it is right or not) or sees "n" behaving like a polynomial is in a strange, distressing, cruel, and arbitrary world where even seemingly familiar things don't make sense, and has every inclination to give up. The presentation of the argument should both make the reading of that argument easy and provide reassurance that all is on course.

Here's a possible analogy to emphasize the difference between argument and presentation. Suppose you were reading a book in which every fourth or fifth word was printed mirror image (so, for example, the first part of this sentence might read "Suppose you erew reading a book ni which every fourth or htfif word was printed mirror egami"). You could surely, yb the dint of noitartnecnoc, decipher each part of eht text. Equally as surely, uoy would read more ylwols, and have much erom difficulty following the actual story enil. Also, you would be annoyed and frustrated at the needless extra work involved in digging out what the person was trying to say.[4] Each of those glitches is a distraction from what ought to be the real task; if you notice the glitch, it is as annoying as somebody kicking the back of your chair while you are taking a test. If you don't notice the glitch, as may happen more plausibly while reading mathematics, you merely struggle on feeling increasingly frustrated and uneasy, unaware that your reading is being sabotaged. The rules above work instead to help the reader be carried along the path of the argument, and that's what is needed. (It is worth remarking that this good writing of proofs has everything to do with writing, and rather little to do with mathematics. Mathematics teachers are sometimes heard to say that it isn't so much the Math SAT that matters as the Verbal SAT, and the vulnerability of mathematical writing to bad writing is the reason why.)

As for specific comments on the rules, few are needed on the first and second, since all they are is fair play for the reader: if you know what you are going to do, and the names you will use, why handicap the reader by making it a secret? But the third rule is more complex and deserves more discussion. Mathematicians attach a lot of importance to these little words, and students frequently do not. But these little words are cues to the *local* structure of the proof: what is going on right now? Well, why does this local structure matter? Here's an example of the first proofs we are likely to get for, say, the result for sets that $A \cap (B \cup C) = (A \cap B) \cup (A \cap C)$:

$x \in A \cap (B \cup C)$

$x \in A$

---

[4]See? Wasn't this last sentence a relief! It isn't a matter of what you can do, it is a matter of what you ought to have to do.

$x \in B \cup C$
If $x \in B$ then $x \in A \cap B$.
If $x \in C$ then $x \in A \cap C$.

Well, there you are. Good luck. All structure has been omitted, and all that a reader has to go on is a list of equations.[5] Even at the level only of "little words," where does the fourth line come from? From the third? The first and third together? Gödel's incompleteness theorem? Who can tell? ... without, of course, essentially redoing the proof yourself, which is what this style of presentation really requires. The sad thing is that the difficulties above are a matter of presentation, because the correct proof (or a piece of it, anyway) is right there — it's just too hard to dig out.

There's a student response to this sort of complaint about little words, and that is to write the proof as shown above and then "throw in" some little words. The advantage is that the proof produced has a certain surface plausibility. The disadvantage, of course, is that the little words can mislead the reader.

Exercise

2.11: What is wrong with the following proof fragment (at the level of "little words")?

Suppose $x \in A \cap (B \cup C)$.
Hence $x \in A$.
Thus $x \in B \cup C$.
Therefore, if $x \in B$ then $x \in A \cap B$.
Or, if $x \in C$ then $x \in A \cap C$.

So to throw in little words that point misleadingly is an admittedly different sin, but just as serious, as their omission altogether. The more the little words are used usefully, the better the reader will follow the local (that is, small-scale) structure of the proof.

The fourth rule, concerning large-scale or global proof structure, is more important than is at first obvious, and we'll say more in Section 2.3 and also when we talk about formal mathematical language in Chapter 3. But, for example, if the proof structure is signaled as "by induction" you ought to have at least a little sensation of relief. Whether or not you remember induction, and especially whether or not you are confident using it yourself, you have seen proofs by induction before and are likely to recognize some similarities in what is to come. In general, standard proof forms help you follow the argument. If, on the other hand, you read that a proof will be

---

[5]At the level of rule 1, did you know, for example, that what was being proved was a containment $A \cap (B \cup C) \subseteq (A \cap B) \cup (A \cap C)$, as half of a strategy that uses the fact that $D \subseteq E$ and $E \subseteq D$ imply $D = E$? How could you know? At the level of rule 2, what is "x"?

by Murcheson's Law of Propinquity, you might well feel daunted. (You should. I just made up both Murcheson and the Law. But the moral is that an unfamiliar form is more likely to be troublesome than a familiar one.)

2.1.5 Exercises

Insert in the following proofs the missing cues to what is going on. A good way to check what you've done is to exchange copies with a classmate. Remember that the goal is not simply to cue *some* proof, but to form and cue a good one that is easy to read.

2.12: The diagonals of a rhombus are perpendicular.
  Proof. $AB \cong CD$. $\angle CAB \cong \angle ACD$ using alternate interior angles. $\angle CDB \cong \angle ABD$ similarly. $\angle DEC \cong \angle AEB$. Triangles AEB and CED congruent. $CE \cong AE$. $DE \cong BE$. $\angle DEC \cong \angle AED$. $\angle DEC$ and $\angle AED$ together form a straight angle. $\angle DEC$ is a right angle.

2.13: A diameter of a circle perpendicular to a chord of that circle bisects it.
  Proof. Let C be the center of the circle, AB the diameter, DF the chord, and E the point of intersection of the chord and the diameter. $AB \perp DF$. Draw CD and CF. $\angle CED \cong \angle CEF$. $CD \cong CF$. $\angle CDE \cong \angle CFE$. Triangle CDE is congruent to triangle CFE. $DE \cong FE$.

These rules of thumb may sound reasonable. As a practical student concentrating on doing some mathematics, though, three questions are fair: Do they help me read proofs? Do they help me discover proofs? Do they help me write up proofs I have discovered? The next section is all about the answer to the first of these and also casts some light on another aspect of these rules.

2.2 Real-Life Proofs vs. Rules of Thumb

The following is an example of the sort of proof one might find in a mathematics textbook at the junior level.

  Prove: The sum of monotone increasing sequences is monotone.
  Pf. If (s_n) and (t_n) are monotone increasing, then of course for any n, $(s+t)_{n+1} = s_{n+1} + t_{n+1} \geq s_n + t_n$, as desired.

Well, ... what happened? If we compare this proof to the guidelines, you could suppose the "as desired" is a signal that the proof has done what it said it would. Only nobody ever said what that was to be. What's "n"? What is "$(s+t)$"? "Then" does seem to point to $(s+t)_{n+1} = s_{n+1} + t_{n+1} \geq s_n + t_n$, but that isn't a single statement. And so on. The rules do not seem to be in force for this proof.

To understand why this proof is (or might be) really OK, we have to go back and think about the rules again and about a standard aspect of (any kind of) writing. First, observe that the rules given have a good deal of redundancy, and, if used completely and slavishly throughout a proof, give *too much* signpost information for a reader. Suppose, for example, I have just told you that we are doing a proof by induction. Should my next sentence be, "since the first part of a proof by induction is a verification for the case $n = 1$, we turn to the case $n = 1$"? Note that this sentence is purely signpost and has nothing to do with the particular proof at hand. Unless I should assume that you have never heard of induction, wouldn't something like "For the case $n = 1$, ..." be good enough? Or, for another example, should every "thus" and "hence" be completely explicit? Do you really want to read many proofs whose form is "Hence, because of A and reason R, we have B. Therefore, because of B and reason S, we have C. Thus, because of C and reason T we have D"? It is surely possible to have even correctly formed signposts so numerous as to create a ridiculously long proof, and, more importantly, to swamp the markers that are really important.[6]

Second, any kind of writing makes assumptions about the *audience* to which it is written. (Those assumptions may be conscious or unconscious, consistent or inconsistent, good or bad, on the part of the writer, but they are always there.) The author of a mathematical proof (or any other argument) has to make some delicate choices, the audience clearly in mind, about how many guides to the structure of the proof and other details are appropriate. A research specialist writing for other research specialists, for example, need only hint at familiar proof structures, may assume immediate access to a wide body of specialized knowledge, can omit as "easy" complex but familiar sorts of arguments, and in general may give a very condensed version of the proof. Someone writing for a novice mathematician, on the other hand, should not. Since we are only concerned at the moment with guides to proof structure, let's analyze what assumptions prevailed in the proof structure above.

Prove: The sum of monotone increasing sequences is monotone.

Pf. If (s_n) and (t_n) are monotone increasing, then of course for any n, $(s + t)_{n+1} = s_{n+1} + t_{n+1} \geq s_n + t_n$, as desired.

I wrote the proof under the assumptions:

1. The definition of monotone increasing sequence and some standard notation for sequences were either immediately recalled or a page turn away for the reader.

---

[6]If you are taking a drive on a new but major highway, do you really need a signpost every mile? Half mile? Tenth of a mile? No, you need enough signposts, whatever "enough" means.

2. Faced with the definition, the reader would quickly deduce that there was only one appropriate proof structure (the one for dealing with the "universal quantifier" on n, whatever that is).

3. Expecting that form of argument, the reader would expect that, because of the details of its structure, an arbitrary *positive integer* would be chosen, and something proved for it.

4. The reader would recognize that, this last task accomplished, there was nothing more to be done.

Now the first assumption has nothing to do with proof structure, but the second, third, and fourth do. If it is assumed that the reader knows or deduces quickly from the problem statement what the form of the proof must be, cues to that structure may be replaced by hints or omitted entirely. If the known proof structure requires the choice of "any positive integer" then "For any n" is probably enough of a clue that the choice is now made, with no more full introduction of this variable name.[7] And a reader familiar with a certain proof structure won't need much recap at the end.

So the proof above wasn't *mathematically* wrong, but its *presentation* may have been wrong for you. This distinction indicates the importance of the fourth rule about familiar proof structures; if you aren't at home with those structures, hints suitable if you were won't be enough to steer you through. We'll turn soon to some familiar structures. But the rules, even in compressed form, are still to aid the reader through the presentation and give guide and comfort for a struggling reader and reassurance for a successful one.

Aside

What about the rules as aids for discovering and writing proofs? They don't help much in discovery, since they present a structure clearly but don't help you find it in the first place. They do sometimes help point out when a proof you think you have discovered isn't right or isn't complete: if you are trying to write one up, come to a "Therefore," and realize that the best you can do is say it points backwards to ... uh ... "all that stuff up there, sort of," your proof is in trouble. The proof structure discussion and

---

[7] Note there is another aid for the reader here, which is custom about variable names. The letters n, m, i, j, and k are, by custom, reserved for integer variables. This custom, suitably extended, and its resulting cues for the reader are so important that we shall henceforth call this rule $2'$, since it is an addition to the rules for the user-friendly use of variables. A variant of this custom is the reservation of e for the identity element of a group, or, in other settings, for the base of the exponential function. You would be unsurprised to find "p" standing for a polynomial and π for the constant ratio of the circumference of a circle to its diameter, and quite surprised to find their roles reversed. Here again is a guide to what is going on that saves the reader needless effort, sparing it for more important things.

the section on formal mathematical language to come will be much more helpful for proof discovery.

As for writing proofs, you won't write them at all well unless you follow these rules consciously or unconsciously. Why not consciously? 'Nuff said. **End Aside**

2.3 Proof Forms for Implication

We start with an observation, which is that most proofs are, or have as the large pieces anyway, a proof of a simple implication whose form is 'if A then B.' Sometimes we write this as 'A implies B,' and sometimes it is even written as 'A only if B' (which is much more confusing; we will discuss in Section 2.3.2 why this makes sense). But as we hinted in Section 1.3, frequently you get an assumption or list of assumptions, and the theorem claims you may deduce a conclusion. For example, you may look at our discussion of the Mean Value Theorem in that section to see this form. For another example, think about some standard theorem in geometry (say, about triangles). Faced with a theorem or subtheorem of this form, there are three approaches that we discuss first.

2.3.1 Implication Forms: Bare Bones

Direct Proof

To Prove	Method	Start Cues
Implication $A \Rightarrow B$ If A then B	Assume A, deduce B	"We prove directly ..." ["Let A be as in the statement"] [None]

End Cues

"B, as desired"
"Q.E.D."
"... as was to be shown"
Restatement of result
[None]

Construct some examples from your past mathematics.

2.14:

Proof by Contraposition
(Indirect Proof)

To Prove	Method	Start Cues
Implication $A \Rightarrow B$ If A then B	Prove 'not $B \Rightarrow$ not A' Assume not B, deduce not A	"We prove the contrapositive" ["We show not B implies not A"] "We precede indirectly" ["Assume (not B)"]

End Cues
"So, by contraposition" Restatement of original '$A \Rightarrow B$' [None]

Give some examples from your mathematical past.

2.15:

Proof by Contradiction
(Reductio Ad Absurdum)

To Prove	Method	Start Cues
Implication $A \Rightarrow B$ If A then B	Assume A and not B, deduce a contradiction	"We prove by contradiction" "Assume, for a contradiction" ["Suppose (not B)"]

End Cues
"We achieve the contradiction" "Contradiction." "#"

2.3.2   Implication Forms: Subtleties

Direct Proof

The logical structure of direct proof is the most simpleminded and most common: assume the hypothesis or hypotheses and deduce the conclusion. Recall that you constructed above some examples from your previous history, which you can compare to the discussion here. Direct proof is frequently cued in the first sentence.[8] Unfortunately, since it is basic and

---

[8]It is legal to use this sentence both to cue the form of the proof as direct and to set up some notation consistent with the hypothesis.

common, sometimes this form is not cued at all; we simply dive into the body of the direct proof, or at least into setting the notation. This is more common if the proof is (under those assumptions about audience) simple or short or easy, but may be done anyway. By convention, if there are no cues, the reader should get to count on a direct proof.

End cues are often left out (or used incorrectly or ambiguously). The "Q.E.D." abbreviating the Latin "Quod Erat Demonstrandum" that you may have seen or written ought to be a reliable cue to the end of a direct proof. It isn't always because writers who ought to know better throw it in because it looks fancy. "As was to be proved" (a reasonable translation of "Quod Erat Demonstrandum") is probably more reliable since anyone using it has at least taken the trouble to think about Q.E.D. and its translation. A phrase like "as was to be shown" or "as desired" is surely the cue to the end of some proof but may not be reserved for direct proofs. The main hope, of course, is that the reader is following the proof well enough to recognize when it is over!

These conventions apply if implication is a subproof in a longer proof, although the beginning and ending of such a subproof are supposed to be very clearly marked. In particular, the cue to the end of the subproof is frequently very explicit: "This completes the proof that under our hypotheses f is uniformly continuous, and we now turn to ...," for example.

Proof by Contraposition

To prove 'A implies B' by contraposition, one makes use of a result from logic that states 'A implies B' is true exactly when 'not B implies not A' is true. (The "contrapositive" of 'A implies B' is 'not B implies not A,' hence the name.) Faced with the task of proving one, it is equivalent to prove the other instead. Thus the proof by contraposition of 'A implies B' amounts to a direct proof that 'not B implies not A.' The beginning of the proof is the assumption of 'not B,' and work to establish 'not A' from it will follow.[9]

The cues for this proof form are several, but at least it is almost always cued. It's not very clear to simply start out "Assume 'not B'," but this is sometimes done. (The reason that it is less clear is that this could well be the start of a proof by contradiction.) The thing to remember is that once you start down this path, the proof is really a direct one, although of something different than what you started with. A subproof of this form might close by restating the 'A implies B' actually needed in the main proof.

---

[9]WARNING: the beginning assumption of 'not B' may have you thinking of "proof by contradiction." Proof by contraposition is not the same as proof by contradiction, although there are some points of similarity. We will get to proof by contradiction shortly.

Proof by Contradiction (Reductio Ad Absurdum)

You probably are familiar with this form (which is perhaps not the same thing as being comfortable with it). To prove 'A implies B,' we begin by assuming 'A' and 'not B.' It is a fact from logic that supposing we can reach a contradiction ('C and not C,' '$0 = 2$,' 'the sum of the angles in the triangle is greater than $180°$,' whatever), we are entitled to deduce 'A implies B.'[10] So the form of the argument is

- Assume both 'A' and 'not B.'

- Arrive at some contradiction.

- Deduce from the first two steps that indeed 'A implies B.'

Examples, please? (We assume that you are merely collecting the ones you constructed after the bare bones discussion. We didn't tell you to do it there, but your habits are so good ... .)

2.16:

This form of proof should always be cued at the beginning and almost always marked at the end. Clear beginning cues are best, although sometimes you will see merely "Assume 'not B' "; as noted before this is ambiguous since this could be the start of a proof by contraposition. In a subproof, a proof by contradiction should be very clearly begun and ended, since you should be sure that a reader is not assuming something later on (and by now known to be false) that was part of the subproof.

There's another, fast and tricky, instance of proof by contradiction you need to recognize. Occasionally you will find a sentence like "Now (something), for if not, then (argument), clearly a contradiction." This is the cue to a small proof by contradiction, usually buried in the middle of a longer argument, short enough to be stated, cued, and argued, all in one sentence. Sometimes the final "clearly a contradiction" is omitted, and then you are on your own to see if this is a one-sentence proof by contradiction, a one-paragraph proof by contradiction, or whatever.[11]

It's worth discussing proofs by contraposition and contradiction, since they are similar enough to be confusing but their differences are important. In each form to prove 'A implies B' you assume 'not B.' In proof by

---

[10]We will discuss this more when we do some formal logic in the next chapter. Intuitively, if you are given 'A,' and assuming 'not B' as well gets you to a contradiction, since the problem can't be with 'A' it must be with 'not B', so 'not B' must be false if we assume 'A', so 'B' must be true if we assume 'A'. This intuitive form is in strict terms nonsense, but may be helpful anyway.

[11]This is not particularly kind to the reader.

contraposition, that is all you assume (besides, of course, general facts from mathematics), and you work more or less directly to 'not A.' In proof by contradiction, on the other hand, along with 'not B' you assume 'A' and now get to work for any contradiction you can find. The obvious advantage of proof by contradiction is that you have *two* assumptions to work with, and somehow having both may give you more things to play with.

There's a convention (part snobbery, part reasonable) about these two that needs discussion. From the discussion above you may realize that one possible contradiction you could arrive at is 'A and not A.' Further, since for this form you are allowed to assume 'A,' all you really need is to get is 'not A.' Since you get to assume 'not B,' this is beginning to look an awful lot like proof by contraposition, and therein lies the problem. Suppose you are reviewing a proof by contradiction and realize that the contradiction was 'A and not A,' and further that the only place the assumption 'A' was used was to pair with the 'not A' deduced along the way for this contradiction. Such a proof could be just as well written as a proof by contraposition: assume only 'not B' in the first place, deduce 'not A' exactly as was done before, and stop. Although the original form as a proof by contradiction is not logically wrong, *it is convention that you write it as a proof by contraposition if you can.* Well, OK, it is indeed more economical not to assume 'A' if you don't need it. But the bad reaction to it goes far beyond that; in some way or other, it is low class to do it the wrong way, even though the proof is correct. (And we wouldn't want to be low class.)

There's another quite sensible convention. There is a general dislike of proofs by contradiction if a direct proof or proof by contraposition is available. We mentioned one reason before: a proof by contradiction shows you why something isn't false, and of course logically it is therefore true. But that doesn't tell you *why* it is true. If there is a direct connection between hypothesis and conclusion, why not show it? Further, on a level below that of logic, we find the argument "it is true because it is not false" unsatisfying because in ordinary living we find plenty of things neither true nor false. Logically the argument is fine, because in mathematics we accept the Law of the Excluded Middle, which does away with anything other than true or false. That still doesn't make it satisfying psychologically. Finally, it is unsatisfying to bend the attention of the reader to matters that we know, and the reader believes, will turn out not to be the case. To assume 'A and not B' and force the reader to hold this in mind for a whole proof is to request the reader to think about a lie. (This point is essentially taken from Pólya's *How to Solve It* [5, page 168]: "... we are obliged to focus our attention all the time upon a false assumption which we should forget and not upon the true theorem which we should retain.") Is it a proof? Of course, and so it is used when there is no alternative. But if a direct proof is available we take it.

Some language needs discussion. We remarked before that 'A only if B' was language used for 'A implies B,' but somewhat confusing language. Its justification is formal logic: the statement 'A implies B' is assigned a value of true or false without regard for anything but the true or false assignments to 'A' and 'B.' (So, in particular, "meaning": "There are no unicorns implies seven is a number" is true.) The convention for the assignment to 'implies' is that it is *false* if 'A' is true but 'B' is false, and 'implies' is true under all other possible combinations of assignments (we'll discuss this more when we do formal language). In particular, if 'A' is true, one must have 'B' true as well, or we would be in the only situation in which 'A implies B' would be false. That is, if 'A' is true, 'B' must be; that is, 'A' *only if* 'B.'

This language occurs again in 'A if and only if B' (sometimes abbreviated to 'A iff B' and sometimes written '$A \Leftrightarrow B$.' This is shorthand for '$(A$ implies B) and (B implies A),' and now that you understand 'only if,' you can see why. The phrase 'A if B' is just 'if B, then A,' and you know what 'A only if B' means. Another phrase for this same logical statement is "double implication." One needs this for the sort of theorem discussed in a footnote in Section 1.8.2. Note that the proof is almost always two subproofs of the two implications hidden in "iff."

There is more alternative language, unfortunately. Sometimes you will see something like "this condition is *necessary* for f to be continuous" or "this condition is *sufficient* for f to be continuous." These are again ways of talking about implications. To say that 'A' is sufficient for 'B' is to say that if you have 'A' you surely will have 'B' (that is, 'A' is enough to guarantee 'B'), which is 'A' implies 'B.' To say that 'A' is necessary for 'B' is to say that you can't have 'B' without having 'A'; if you think of the truth table for 'implies,' you will see that this is '$B \Rightarrow A$.' Of course, these can be combined, so one can talk about a necessary and sufficient condition, which is really an 'if and only if' in disguise. The fact that there are so many ways to say the same thing is confusing, but they are in common use so you have to know them.

Sometimes one has a theorem in which one is showing that several (more than two) conditions all imply each other. The language for this is "the following are equivalent," sometimes abbreviated to "TFAE"; if your conditions are 'A,' 'B,' 'C,' and so on, this is short hand for '$(A$ iff B) and $(A$ iff C) and $(C$ iff B),' and so on, and is a great convenience. It is worth noting that to prove this, it is logically enough to prove the circle of implications 'A implies B,' 'B implies C,' ..., 'M implies A.' This cuts down the number of implications to be proved from $n(n-1)/2$ to n, a substantial savings.

2.3.3 Exercises

In the following you are to identify the proof structure, critique the cuing of that structure, and note any proof done by contradiction that should have been done by contraposition.[12] A first pass might be to highlight *all* cuing words.

2.17: Suppose that f and g are functions for which the composition $h = g \circ f$ is defined. If h is injective, then f is injective.

Proof. Suppose that f is not injective. By definition, then, there are distinct points x_1 and x_2 in the domain of f so that $f(x_1) = f(x_2)$. Then clearly $h(x_1) = (g \circ f)(x_1) = g(f(x_1)) = g(f(x_2)) = h(x_2)$, and therefore h is not injective.

2.18: The identity element in a group is unique.

Proof. Let G be a group and suppose e and f act as identity elements. Then $e = e * f$ since f is an identity, and $e * f = f$ since e is an identity. Combining these equations, $e = f$ and so these elements are the same.

2.19: The identity element in a group is unique.

Proof. Let G be a group and suppose e and f are distinct elements that act as identities. Then $e = e * f$ since f is an identity, and $e * f = f$ since e is an identity. Combining these equations, $e = f$, a contradiction.

2.20: Let $\{A_\alpha\}$ be a collection of connected sets with nonempty intersection. Then $\cap_\alpha A_\alpha$ is connected.

Proof. Let p be a point in the intersection, and suppose that $\cap_\alpha A_\alpha$ is not connected. Then there exists a disconnection C, D as usual; in particular, neither C nor D is empty and $\cap_\alpha A_\alpha = C \cup D$. Without loss of generality, we may assume that $p \in C$. Now for each α, $p \in A_\alpha$, and therefore $C \cap A_\alpha$ is not empty. Therefore, since A_α is connected, A_α is contained entirely in C. It follows that D is empty, which is a contradiction.

2.21: Let S be a set of real numbers bounded above, and denote by $-S$ the set $-S = \{-x : x \in S\}$. Then $-S$ is bounded below.

Proof. Let b be an upper bound for S. For any y in $-S$, there is an x in S such that $y = -x$. Further, $x \le b$ since b is an upper bound for S, and so of course $-b \le -x$. Thus $-b \le y$, and it follows that $-b$ is a lower bound for $-S$.

2.22: Let $\{v_j\}_{j=1}^n$ be a linearly dependent set of vectors in a vector space V (with $n \ge 2$). Then there is a subset of the v_j with $n - 1$ elements with the same span as the original $\{v_j\}_{j=1}^n$.

Proof. We construct the required subset. Since the set $\{v_j\}_{j=1}^n$ is linearly dependent, there exist scalars $\{\lambda_j\}_{j=1}^n$, not all zero, so that $\lambda_1 v_1 + \lambda_2 v_2 + \ldots + \lambda_n v_n = 0$. Without loss of generality, we may assume that $\lambda_1 \ne 0$. We then claim that the set $\{v_j\}_{j=2}^n$ has the required properties, and clearly the number of elements is correct. To show that its span is the same as that of the full set, suppose v is in the span of $\{v_j\}_{j=1}^n$, so there exist scalars

---

[12]This last task depends on Subsection 2.3.2.

$\{\alpha_j\}_{j=1}^n$ such that $v = \alpha_1 v_1 + \alpha_2 v_2 + \ldots + \alpha_n v_n$. From the dependence equation, we have $v_1 = -\lambda_2/\lambda_1 v_2 - \ldots - \lambda_n/\lambda_1 v_n$, and it is easy to deduce from the last two equations that v is in the span of the $\{v_j\}_{j=2}^n$. We have therefore constructed a set with the desired properties.

2.23: Cographs of isomorphic graphs are isomorphic.

Proof. Let G and H be graphs, and h an isomorphism between them. By definition, we must show that there exists an isomorphism between G^c and H^c; we will show that h is such an isomorphism. Surely h is a one-to-one function from the vertex set of G^c onto that of H^c, since the vertex sets of G and G^c, and H and H^c, respectively, are the same. We next must show that for any two vertices v_1 and v_2 of G^c, the edge v_1–v_2 is in G^c if and only if the edge $h(v_1)$–$h(v_2)$ is in H^c. Suppose v_1–v_2 is in G^c; then by definition, v_1–v_2 is not in G. Since h is an isomorphism of G and H, $h(v_1)$–$h(v_2)$ is not in H. By definition, $h(v_1)$–$h(v_2)$ is in H^c, as desired. The other implication is similar, and we are done.

2.24: If a is an element of a group such that $a*a = a$, then a is the identity element.

Proof. Let e denote the identity of the group. There exists b such that $b * a = e$. Observe $b * (a * a) = e$ from our assumption on a. Also, $(b * a) * a = a$. Therefore, using associativity, we have $a = e$ as desired.

2.25: Suppose that $a_1 + \ldots + a_9 = 90$, with the a_i nonnegative integers. Then there exist three of the a_i whose sum is greater than or equal to 30.

Proof. WLOG, we may suppose the a_i are in decreasing order. If $a_1 + a_2 + a_3 \geq 30$, we are done. If not, then clearly $a_4 + a_5 + a_6 < 30$ and $a_7 + a_8 + a_9 < 30$, a contradiction.

2.26: Suppose f and g are functions for which the composition $g \circ f$ makes sense. If f and g are injective, then $h = g \circ f$ is injective.

Proof. We use the definition of injective directly. Suppose that x_1 and x_2 are elements such that $h(x_1) = h(x_2)$. Then by the definition of h, $g(f(x_1)) = g(f(x_2))$. Since g is injective, we may deduce $f(x_1) = f(x_2)$, and since f is injective, we may deduce $x_1 = x_2$, as desired.

2.27: The intersection of an arbitrary nonempty family of subgroups of a group is again a subgroup.

Proof. We use the theorem stating that a nonempty subset S of G is a subgroup if and only if for every a and b in S, $a * b^{-1}$ is in S. Let $\mathcal{S}$ denote the family of subgroups and T denote the intersection. To show T nonempty, note that since the identity element of G is in S for each $S \in \mathcal{S}$ it is surely in T. Suppose a, b are in T. Then a is in S for each $S \in \mathcal{S}$ and similarly for b. So for each S, $a * b^{-1} \in S$ citing the theorem. Then $a * b^{-1} \in T$ from the definition of intersection, and we are done via the theorem.

2.28: Let G be a graph, and define a relation R on the vertex set $V(G)$ by $(a, b) \in R$ if and only if there is a walk from a to b. Then R is an equivalence relation.

Proof. We verify the three conditions for an equivalence relation. First, there is the trivial walk from any vertex to itself, so $(a, a) \in R$ for each a. Second, suppose there is a walk from a to b. By reversing the order of the list of vertices, we may obviously produce a walk from b to a, which is symmetry. Finally, suppose (a, b) and (b, c) are in R. By simply concatenating the lists of vertices for the two walks, we may obviously produce a walk from a to c, which is transitivity. Thus we are done.

2.29: If G is a group and a is any fixed element of G, we let $T_a : G \to G$ denote the function of right translation by a, so $T_a(x) = x * a$ for all $x \in G$. Prove that T_a is injective.

Proof. We use the definition of injective, so suppose $T_a(x) = T_a(y)$. Then $x * a = y * a$. Multiplying on the right of each side of this equation by a^{-1} it is easy to deduce $x = y$ as required.

2.3.4 *Choosing a Form for Implication*

We hope that the above discussion helps with two of the three standard questions, namely, "does it help me read proofs?" and "does it help me write up proofs I have found?"[13] We still need to say a few words about choosing a proof form during your attempts to discover a proof. Such a choice is clearly important, since, for example, if you decide on a direct proof form and there is no direct proof but only a proof by contradiction, your work and time may not be entirely wasted but won't yield success. To choose a certain form is to plan on making an investment of time and energy; how do you know which form is best?

The answer is that you don't, and anyone trying to give you some perfect general recipe is either deluded or lying. Frequently you have to try all three forms, perhaps starting with the direct form since it is somewhat preferable. Often your choice is guided by "experience," which can sometimes be reduced to rules and often can't. Surely everything you learned in the chapter about looking at examples should be used, and we will come to some formal mathematical things in the next chapter helpful in some cases. The best we can do is give some rules of thumb that might increase the chances of your choosing a successful form. We'll need a little discussion first.

What is the discovery? of a proof One way to think of things is to visualize the hypothesis as one point ("H") and the conclusion as some other point ("C"), with the job being to connect the two. The connection is usually done with the aid of some intermediate points, so, for example, you find

---

[13]We hope each answer is a resounding YES! If not, keep it to yourself.

some connection from "H" to "A", and some connection from "A" to "B," and some connection from "B" to "C," and you are done. Note that while you had "H" and "C" to start with, you had to choose "A" and "B" from all the other possibilities in your mathematical repertoire, most of which are totally irrelevant to this problem. How do you do that? Well, a good place to start is by seeing what "H" connects to naturally. Perhaps there are three theorems available connecting "H" to "A," "D," and "E." And, since as we will argue later, connections *to* "C" are at least as important, perhaps there are four theorems connecting "B," "R," "S," and "T" to "C." This gives a picture something like this:

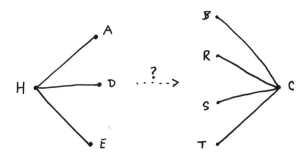

Now, of course, you have the same sort of problem over again, because you want to find a connection from "A" to any one of "B," "R," "S," and "T," or a connection from "D" to any one of "B," "R," "S," and "T," or ... . New problem: since you can't think about all of these at the same time, you have to concentrate on one. Which one, and how do you know? Well, that's the hard part, and a full discussion would take us far afield.[14] But if somehow you are able to find a connection from "A" to "B" (or, for that matter, from "D" to "R") you win.

How does this model help us think about choosing a proof form? If when faced with 'A implies B' you perceive many strands leading from 'A' and/or many strands leading to 'B,' it is probably worth trying a direct proof. You might expect this, for example, if your hypothesis contains a nice, juicy, technical definition, or perhaps if your hypothesis has a number of different conditions. Pólya, in *How to Solve It*, lists a number of questions designed to aid in the discovery of these strands (for example [5, page 156], "Could you derive something useful from the hypothesis?"). Start by examining these strands.

If that doesn't seem to go very far, what if there are many strands leading from 'not B' and/or many leading to 'not A'; what kind of proof might you choose?

---

[14]A very abbreviated discussion of this, but with further references, can be found in the Theoretical Apologia appendix and in the final section of Chapter 3.

2.30:

Finally, if the things above fail, there is always proof by contradiction. But you might be led to this form earlier, either by apparent lack of usefulness of the hypothesis or by something special cue. Consider the Goldbach conjecture (still unresolved) stating that every even number is the sum of two or fewer prime numbers. A little work makes you realize that this is really saying that every even number larger than six is the sum of exactly two (odd) prime numbers. So you are faced with some number, even and larger than six. What do you know about such numbers? Not much apparently useful, so perhaps it is time for proof by contradiction?[15]

There is a special cue to proof by contradiction as well. One form of that cue is a 'not' in the conclusion, so if you are asked to prove "if (blah blah blah) then (something) is *not* (something else)" the standing of a proof by contradiction goes way up. To show something is not a circle, for example, you could either show it *is* one of the rather larger number of other things (ellipses, cows, catenary curves, almost periodic matricial functions, ...) or you could assume it is a circle and deduce a contradiction. Another way to say it is that "not circle" is too general to be useful, but "not not circle" (that is, "circle") is extremely useful.

There's another way you can be in the same situation but have it less noticeable. A few definitions in mathematics have a 'not' already in them. For example, in analysis there are two sorts of sequences one singles out for study, one being the convergent sequences and the other being the sequences divergent to plus or minus infinity. Without worrying about what these are, it should be unsurprising that there are sequences in neither of these two categories. So one makes up a third category, called oscillating sequences, which consists of everything else: that is, a sequence is oscillating if it is not convergent and not divergent to plus or minus infinity. (In some sense, these are the ones you'd like to sweep under the rug.) Now, how will you prove that some sequence is oscillating? Oscillating is not a positive property — you can't point to something the sequence does and say "See? It does that so it is oscillating." All you can do is a proof by contradiction. Assume, you say, that the sequence is not oscillating. Then it is either convergent or divergent. ... Contradiction(s), so the sequence is oscillating.[16]

---

[15]That hasn't worked either (yet?), but it's a reasonable try.

[16]We mention that there is in topology a similar definition. A set is said to be connected if it is not disconnected. A set is disconnected if it actually does (something), and assuming a set is disconnected gives you some things to work with. So to prove a set is connected you almost always proceed by contradiction, assuming for a contradiction that it is not connected, that is, not not disconnected, that is, disconnected. See what fun things lie ahead?

These rules of thumb notwithstanding, about all you can do is improve your chances of picking a suitable proof form. The more proofs you do, the better your choices will become. And at least as important as the initial pick is your ability to monitor what you are doing so you pull your head out of the sand if your pick isn't being useful.[17] Try several proof forms, keep your wits about you, and see what you get.

2.4   Two More Proof Forms

We single out two more forms, proof by cases and proof by induction.

2.4.1   Proof by Cases: Bare Bones

Sometimes in a proof you find yourself needing to prove some conclusion "C" under two (or more) different sets of circumstances. Suppose, for example, that you need to prove that the square of a nonzero number n is positive. A natural argument divides things into two cases:

Case I. $n > 0$ Then $n^2 = n * n > 0$ since the product of positive numbers is positive.

Case II. $n < 0$ Then $-n > 0$, and $n^2 = n * n = 1 * (n * n) = (-1 * -1) * (n * n) = (-n) * (-n) > 0$, again using the fact that the product of positive numbers is positive.

The example is, of course, trivial, but the argument form is important. You are really faced with proving something of the form '(P or Q) implies R,' and you do it using cases, in the first of which you assume 'P' and in the second of which you assume 'Q.' Since in each case you arrive satisfactorily at 'R,' you do indeed get what you want.

Proof by Cases

To Prove	Method	Start Cues
$(P \text{ or } Q) \Rightarrow R$	Prove '$P \Rightarrow R$,' then prove '$Q \Rightarrow R$'	"We prove by cases ..." ["To show $P \Rightarrow R$"] ["If P, then ..."]

Case Start	Case End	End Cues
"Case 1 (2, ...)" "In the first case"	"...finishes Case 1" "So done in first case"	"So in either case" [None]

This form is useful and common but needs to be well cued, since, for

---

[17]There is a class of exercises which work on exactly this skill; see the discussion in Section 3.5.

example, while in the middle of case II you do not get to use the assumption of case I. Note that there are six places to mark: the beginning of an argument by cases, the beginning and end of case I, the beginning and end of case II, and the end of the argument by cases as a whole. It is probably always worth tying things neatly up at the end.

2.4.2   Proof by Cases: Subtleties

There is, out there, an inadequate and truly annoying but not unused method of dealing with proof by cases. Occasionally you will read a proof that has early on something like "Assume $A \neq B$." You may at the time see no particular reason why this is a legal assumption, and your efforts to see why you can in fact make it may get you nowhere. Possible cure: skim your way down the rest of the proof. If you come upon a paragraph starting "In the case $A = B$, ..." then you have in fact been in the first case of a badly cued argument by cases (you ought to be allowed to shoot the author, but this is not practical). The best you can do is be alert for this use of "assume" ("suppose" is sometimes used in the same way).[18]

Let's warn you about a common mathematical device, or perhaps devices, for dealing with arguments by cases, essentially by omitting one of the cases. One of the favorite phrases of mathematicians is "**without loss of generality**" (sometimes abbreviated to **WLOG** or **W.L.O.G.** or **wlog** or **w.l.o.g.**), and you may have wondered what it meant. Well, suppose, for example, that you are concerned with some number a in a set A and another number b in a set B, and sets A and B are not the same but the properties you care about for this problem are the same (for example, they are both open intervals). You may see something like "$a = b$ is trivial, so WLOG, we may assume $a < b$," and you may well wonder; it certainly seems that $b < a$ is possible, so we are losing some "generality." We are, but what the WLOG is signaling is some argument like this: surely there are indeed two cases, $a < b$ and $b < a$. But the argument I am going to give for the $a < b$ case is exactly the argument for the $b < a$ case, except I would have to change each a to a b and each b to an a the second time around. So I won't bother to do that, but I'll just remind you by saying "WLOG" that this is going on and leave you to fill in the details if you feel it necessary.

---

[18]We pause to note that the case for good recognition of proof forms has just received another bit of support. We have, so far, observed that "assume" could be a cue to a proof by contradiction, a proof by contraposition, or a proof by cases. If you have these three forms well in mind, what follows the "assume" will probably let you choose among them. But if the forms aren't readily available you may not get expectations or, worse, get incorrect ones. Remember, a word to the wise ... .

Often this signal is used when the division into cases really resulted solely from some notational choices. For example, what we really had were two sets and an element from each set. If we could look ahead and see which element was smaller, we could agree to name the lesser one a and call its set A, and the greater one b and call its set B; and then the two cases would never have arisen. This is the use of "WLOG" requiring the least cuing, since it is a frequent one. Sometimes, though, "WLOG" is stretched to other situations in which the two cases don't arise simply from the notational choices, and sometimes even aren't quite the same. Usually in that case there is some reason given along with the "WLOG" signal to say why the two cases aren't really enough different to worry about; for example, "without loss of generality we may assume f is positive, since if not, apply the argument to $-f$." Here we have a guarantee that in the judgment of the author the second case is sufficiently like the first case so that, with this hint, the reader will be able to complete the proof of the second case. (Remember the matter of audience assumptions? Here is a good example of such an assumption.) But however stretched, WLOG is always a cue to some case about which you will hear no further, because the proof of that case will be omitted. An example of the use of WLOG can be found in Exercise 2.22 above. This sort of division into cases (one to be omitted) may also be cued by "assume," but sometimes this cue is strengthened a little by saying, "Assume, say, that $a < b$." The "say" is to point out that indeed a choice is being made, but to hint that the writer or you will soon note that it doesn't make any real difference.

Another device for omission of cases is to prove one case completely and then say "the other case follows *mutatis mutandis*." The phrase *mutatis mutandis* is a Latin one and may be roughly translated as "changing that which needs to be changed." Here's another guarantee on the part of the author that the reader will be able to complete the omitted case by using the argument from the first case with a few simple(?!) changes. Finally, a last way to dispose of one kind of case is to say "except for the trivial case (something), ... "; here's a guarantee that there is an omitted case, but in the opinion of the author any idiot is supposed to see why things are true in that case. But realize that in all these devices a shortened argument by cases is still there.

It should finally be pointed out that the discussion above behaves as if the only arguments by cases are those with two cases. It is a reasonable test of your understanding to take the time to write down what would happen if there were three cases, for example. And, of course, you might try to find, construct, or recall a concrete example of such an argument.

2.4.3 Proof by Induction

To complete the standard proof forms we ought to discuss induction. One place to start is with what you may already have learned, which is that to prove something by induction you first prove it for the case $n = 1$ and then, assuming its truth for n, prove it for $n + 1$.[19]

Proof by Induction

To Prove	Method	Start Cues
For all n, $P(n)$ (formula involving n for all positive integers n)	Prove $P(1)$, then prove for all n, $P(n) \Rightarrow P(n+1)$	"We prove by induction ..."

"$n = 1$" Start	"$n = 1$" End	End Cues
"For $n = 1$"	"...finishes $n = 1$ step"	"Done by induction"

"Induction Step" Start	"Induction Step" End
"For the induction step" ["Suppose $P(n)$"]	"Induction step is done" ["So $P(n+1)$ as desired"]

You've seen this most in proofs of equalities true for all n, like $\sum_{k=1}^{n} k = n(n+1)/2$, but induction has many other uses. Luckily, the cuing is quite standard, as indicated above.

2.4.4 Proof by Induction: Subtleties

There are a few subtleties about induction. First, we have presented things above as if the "first" case is always for $n = 1$. It need not be; much less common but still occurring are $n = 0$ and $n = 2$, and in fact, if one wants to prove something for the sequence of integers $\{k, k + 1, k + 2, \ldots\}$ then the starting value is k.

Second, there is another form of induction, sometimes called "weak" induction, in which the induction step is different. Instead of assuming the statement holds for n and proving it holds for $n + 1$, one assumes it holds for all integers k between 1 and n, and proves that the statement holds for $n + 1$. This clearly gives you more to work with and sometimes is useful. Other than the announcement at the beginning that "weak" induction is being used, the cuing is the same.[20]

2.4.5 Exercises

With the various proof forms in mind, please determine the structure of these proofs and of subproofs if they occur. One way to show the structure might be to shade or highlight sentences, so, for example, a block forming

---

[19]There is implicit here the assumption that n is a positive integer.

[20]These statements about induction are rather imprecise. When we do formal language in Chapter 3 we can say things better.

a subproof by contradiction inside a proof by cases would be all the same color; it might be helpful as well to highlight cuing words. "It might be helpful" because the second task, after the proof structure is clear, is to critique the cuing: was it enough? too much? ambiguous? And, finally, are there any proofs by contradiction that should have been written as proofs by contraposition?

2.31: For any sets A, B, and C, $(A \cap C) \cup (B \cap C) \subseteq (A \cup B) \cap C$.

Proof. Let x be an arbitrary element of $(A \cap C) \cup (B \cap C)$. If $x \in A \cap C$, we have $x \in A$ and $x \in C$. Then surely $x \in A \cup B$ and $x \in C$, and therefore $x \in (A \cup B) \cap C$. The other case is similar.

2.32: Suppose d is a metric on a space M, and we define d_b by $d_b(x, y) = \min(d(x, y), 1)$. Then d_b is a metric on M.

Proof. We must check the three conditions for a metric. Since $d_b(x, y)$ is the minimum of nonnegative quantities it is nonnegative, and it can only equal zero if $d(x, y) = 0$, which implies $x = y$ as needed. Also, $d_b(x, y) = \min(d(x, y), 1) = \min(d(y, x), 1) = d_b(y, x)$ since d is a metric. For the third, we must show that for any x, y, and z,

$$d_b(x, y) \leq d_b(x, z) + d_b(z, y).$$

If either $d_b(x, z)$ or $d_b(z, y)$ is one, the inequality is obvious, so we turn to the case $d_b(x, z) < 1$ and $d_b(z, y) < 1$. Now $d(x, y) \leq d(x, z) + d(z, y)$ since d is a metric. Then from this and our assumption, we have $d(x, y) \leq d_b(x, z) + d_b(z, y)$. Combining this with $d_b(x, y) \leq d(x, y)$, we have the result desired in this case as well.

2.33: For any $n \geq 1$, the sequence n, n, $n - 1$, $n - 1$, ..., 2, 2, 1, 1 is graphic. (See Exercise 1.92 for the needed definition.)

Proof. It is clear that the sequence 1, 1 is graphic: just connect two vertices by an edge. Suppose that for all j less than or equal to n, the sequence j, j, $j - 1$, $j - 1$, ..., 2, 2, 1, 1 is graphic; we must show that $n + 1$, $n + 1$, n, n, $n - 1$, $n - 1$, ..., 2, 2, 1, 1 is graphic. Since $n - 1$, $n - 1$, ..., 2, 2, 1, 1 is graphic, there exists some graph G for which it is the degree sequence. We shall construct the desired graph H by adjoining some vertices and edges to G. Collect the vertices of G into two subsets A and B, each of which has one vertex of degree $n - 1$, one of degree $n - 2$, and so on down to 1. We add 4 vertices to G to produce H. Add vertices x_1 and y_1. Add two more vertices X_1 and Y_1; connect by edges X_1 to every vertex in A and to x_1, Y_1 to every vertex in B and to y_1; and put an edge between X_1 and Y_1. It is easy to check that the resulting graph has the required degree sequence.

2.34: Definition: for x a real number, we define $|x|$ by $|x| = x$ if $x \geq 0$ and $|x| = -x$ if $x < 0$. Prove that for any x, $|-x| = |x|$.

Proof. If $x = 0$ the result is trivial. If $x > 0$, then $-x < 0$, so by definition $|-x| = -(-x) = x = |x|$. If $x < 0$, then $-x > 0$, and $|-x| = -x = |x|$ again by definition.

2.35: Suppose L is a collection of n straight lines in the plane, no two parallel and with $n \geq 2$, and such that no more than two lines meet at any point. Then there are $n(n-1)/2$ points of intersection of the lines in L.

Proof. It is trivial to verify the formula for $n = 2$. Suppose the formula holds for some $k > 2$. To show it holds for $k + 1$, remove one line from the collection. It is trivial to verify that what remains is a collection of k lines, no two parallel, and such that no more than two lines meet at any point. By hypothesis, then, there are $k(k-1)/2$ points of intersection of this smaller collection of lines. Now reintroduce the deleted line. Since it is parallel to none of the k others, it has k points of intersection with them, and none of these has already been counted since in that case there would be a point on three lines, disallowed. So there are $k(k-1)/2 + k$ points of intersection, and a little algebra completes the result.

2.36: Suppose M is a metric space with metric ρ. Suppose A is a dense subset of B and B is a dense subset of C. Show that A is a dense subset of C.

Proof. We use the definition: recall that X is a dense subset of Y if for every $y \in Y$ and every $\epsilon > 0$ there exists $x \in X$ so that $\rho(x, y) < \epsilon$. Therefore we must show that for every point c of C and every $\epsilon > 0$ there exists a point a of A such that $\rho(a, c) < \epsilon$. So let c and $\epsilon > 0$ be arbitrary. Since B is dense in C, using the definition applied to c and $\epsilon/2$ there exists $b \in B$ such that $\rho(b, c) < \epsilon/2$. And since A is dense in B, using the definition applied to b and $\epsilon/2$ there exists a in A so that $\rho(a, b) < \epsilon/2$. It is then easy by the triangle inequality to deduce that $\rho(a, c) < \epsilon$, so a is the point desired. Since c was arbitrary in C and $\epsilon > 0$ was arbitrary, we are done.

2.37: Prove that in any group G, $(a * b)^{-1} = b^{-1} * a^{-1}$.

Proof. Observe that $(a * b) * (b^{-1} * a^{-1}) = e$ by an easy application of associativity. Therefore $b^{-1} * a^{-1}$ is an inverse of $a * b$. Since inverses are unique, it must be "the" inverse $(a * b)^{-1}$.

2.5   The Other Shoe, and Propaganda

Why the sad face? Well, we've finally run smack into "quantifiers," whatever they are.[21] Because of lack of correct quantification, the above description of proof by induction is nonsense if taken literally and vague at exactly the important point if taken figuratively. Further, in the proofs in the preceding exercises there was a lot of quantification going on, cued and uncued, which you may have missed entirely. Surely if the structure of the proofs you observed didn't include some quantification, there was a whole layer of things you didn't see. So a discussion of quantification is the next order of business.

It may prove surprising that the discussion will be the next chapter and not just another section, and also that the discussion will be much more formal than that we've just finished. You are hereby given fair warning that this is the author's decision. What follows is the reason for that decision.

The proof structures so far discussed (with the exception of induction) are familiar both from mathematics and from ordinary argument. Their cuing needs to be learned, and, of course, choosing a proof form during discovery is genuinely hard, but the structures themselves are not so difficult. In particular, they are probably learned or half-learned by observation and osmosis from past mathematics classes, high school geometry in particular (audience assumption!) On the other hand, proofs involving quantification, while not necessarily harder, are for many students much less familiar or completely unfamiliar. While quantification has always been there in mathematics, it is often handled in ways transparent to the student, particularly in high school and even continuing into calculus. One may well survive mathematics until about the sophomore year in college without ever having to worry about quantification. Further, nonmathematical arguments

---

[21]Quite genuinely, because the author could figure out no way to stall any longer.

don't often need care in quantification or, often, need quantification at all. The author therefore believes, and experience seems to corroborate this belief, that the attempt to teach proofs involving quantification by providing teacher models and hoping for adequate student imitation is doomed to failure for *many or most* students. Fortunately or unfortunately, though, it is impossible to flourish and difficult to survive in abstract upper-level college mathematics without being able to handle these structures. In the next chapter we turn therefore to an explicit, and substantially more formal, discussion of this level of proof structure.

3

Formal Language and Proof

3.1 Propaganda

We finally turn our attention to quantifiers, which we will discuss at rather a formal level. But the use of quantifiers and proofs involving them is part of the larger discussion about why mathematical proofs are hard to read or write. The reason is that mathematical proofs (at least those written in paragraph form) are written in a mixture of two languages. The last chapter was all about the first language, which is essentially bits of ordinary English borrowed to produce the paragraph; frequently this part of the language serves to *indicate* the structure of the proof. But the structure of the proof is *dictated* by the second language, which is pieces of formal logic borrowed and adapted for mathematics.[1] A good deal of the difficulty with writing mathematics is that most people don't know the formal language of mathematical logic, and so have to pick up the borrowed parts bit by bit. It's a little as if you were trying to learn how to write in a mixture of French and Spanish, and, while you were fluent in French, all you knew about Spanish was the bits you had seen other people write in the mixture, but people expected you to write as if you were fluent in Spanish as well.

One solution would be simply to take a course in mathematical logic, but that's a large investment in time, and most mathematicians don't do it. What we will do here is to try to give you enough grounding in logic

---

[1] If you think about it, the discussion of proof structures in Section 2.3 was really about the easy part of this formal language.

so that you can work with quantifiers in proofs. Along the way we will see some overlap of mathematical logic (the formal language) with the informal and be able to state precisely some things we rather talked around in the previous chapter. The goal is not to make you a mathematical logician; the goal is to make you *comfortable enough* with quantifiers.

Why does anyone bother with formal logical language anyway? It has numerous disadvantages: it is very formal, with tightly prescribed rules of grammar and manipulation. It is quite limited, with a rather small vocabulary. The things it can talk about form a very small part of the possible objects of human thought and so it is nowhere near as rich as a natural language such as English, Japanese, or Bantu (you can't, for example, talk about poetry, love, or the weather). But the payoff for the restrictions and high degree of structure is great efficiency (a mathematical sentence may take pages of English to translate) and great precision (lack of ambiguity). That precision is really required to do mathematics.

With this going for it, why is formal mathematical language feared and loathed by so many students, particularly those who are having to grapple with writing proofs? The flip side of "efficient" is "intimidating, densely written, and hard to read"; the tightly prescribed "rules" are easy to violate, which makes it hard to write.[2] The language of mathematics just doesn't seem very user-friendly, and, since you get to see only its "bits and pieces," nobody gives you a fair chance to learn it in the first place.

The above negatives are partly true, but the requirement for precision overrides them. The goal of this chapter therefore is, first, to give you a fair chance to learn formal mathematical language and, second, to convince you that a sensitivity to its use can make your life as a prover of mathematics much easier. In particular, the way it is written can help you discover proofs, if only you learn the clues hidden in the writing.

Let's give an example of how sensitivity to formal structure is useful. Consider the following:

$$5 = 4 <> 5 + .(+ \geq \sqrt{,} , = .$$

You see instantly that this mish-mash makes no sense for "grammatical" or "structural" reasons. The problem isn't the meaning of the symbols ("5" still means what it always does), it is that they aren't put together correctly. For a more positive example, consider the following incomplete expression:

$$f(x) = a_{17}x^{17} + a_{16}x^{16} + \ldots$$

which you are indeed likely to think incomplete. What might come after the ...? Lots of things might, but you unconsciously rule out many others. Your familiarity with the language of mathematics gives you an expectation

---

[2]There's always a professor waiting to pounce on the slightest error as if it had earthshaking consequences.

of what might come next, and that expectation is likely to be helpful. (Suppose that thing after the ... was $\lim_{x \to 14.53} x^2$. Would you be surprised? Hope so!) The next sections are to raise your awareness of other linguistic structures in more advanced mathematics and point out the benefits of such awareness.

3.2   Formal Language: Basics

We begin with a brief discussion of the basics of symbolic logic. Since this isn't a text on logic, we will indeed be sketchy and as informal as possible. The aim is to give you the sort of rough and ready understanding of these matters working mathematicians use in proving things, not the precise and formal system a logician might desire.

Recall that the basic building block of logic is the statement: a sentence either true or false. These are frequently abbreviated by single letters, so 'P' might stand for "It is raining." Recall also the connectives that allow one to build compound statements: 'and,' 'or,' 'not,' (denoted $\neg$) and 'implication' (denoted $\Rightarrow$). For each assignment of true or false to statements used in a compound statement, the compound statement has a resulting assignment of true or false. For example, if 'P' is assigned true then '$\neg P$' is assigned false, and if 'P' is assigned false then '$\neg P$' is assigned true. For the more complicated compound statements, things are most easily summarized in truth tables, the basic ones shown below:

P	Q	P and Q
T	T	T
T	F	F
F	T	F
F	F	F

P	Q	P or Q
T	T	T
T	F	T
F	T	T
F	F	F

P	Q	$P \Rightarrow Q$
T	T	T
T	F	F
F	T	T
F	F	T

Let us call attention to the table for $\Rightarrow$, which has the surprising feature that if the first term is assigned false, the statement of implication is automatically assigned true. Also, the truth table for 'or' indicates that this is the inclusive "or," in which we agree to call the compound 'P or Q' true even if both P and Q are true. Frequently in ordinary English we have in mind the exclusive 'or': I'll go to the movies or I'll go to the beach. If I do both, you might say I lied; in mathematics we agree to allow that.

We mentioned in a previous section, but stress again, that these assignments of true or false to compound statements are completely mechanical (based on the truth values of the components and the truth tables that give the rules), and have nothing to do with the *meanings* of the components or whether the compound statement is sensible in ordinary terms. For example, "It is snowing and grass is green" is true if it is in fact snowing. Well, that's not too bad, but it is hard to swallow that "dolphins eat sausage implies the sky is blue" is true, but it is (check the truth table, and some sky in your neighborhood). Since logic was developed as a way to "reason mechanically," the labels "true" and "false" are helpful *when* we want to use logic to reason with. But from a more formal point of view, the words "true" and "false" are misleading, since they make us think of ordinary truth and falsehood; all of logic could do as well on formal grounds if we used the labels 'A' and 'B' instead. We won't get this formal but warn you to distrust your intuition a little.

In reading and writing compound statements one must either use a great many parentheses or have some notational conventions. We'll simply have forests of parentheses if required.

Recall that some statements, such as 'P or $\neg P$,' are always true, independent of the truth assignment to 'P'; such a statement is called a tautology. Check that 'P or $\neg P$' is a tautology by a truth table, just for practice:

3.1:

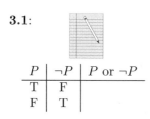

P	$\neg P$	P or $\neg P$
T	F	
F	T	

Some pairs of statements are either both true together or both false together under any possible assignment of true and false to their component statements. For example,

$$P \text{ and } (Q \text{ or } R) \qquad \text{and} \qquad (P \text{ and } Q) \text{ or } (P \text{ and } R)$$

form such a pair. Such a pair of statements is called equivalent, and it is frequently necessary in a proof to exchange one statement for an equivalent one. Check that the above statements are indeed equivalent, by means of a truth table.

3.2:

How do you prove things in this system? Well, we first have to decide what a proof is, and we'll take a proof to be a list of statements, the last of which is the conclusion desired. We will sometimes include justifications for the statements, yielding something like this:

1. Statement 1   Reason 1
2. Statement 2   Reason 2
    . . .

What may these statements be? We allow assumptions, steps inferred from previous steps, tautologies, axioms, definitions, and previously proved theorems. It is worth noting that the proof *is* the list of statements; the "reasons" are a commentary on the proof (although this commentary is often the only thing that makes the proof readable). In particular, we signal the use of an deduction form (a logic professor would say inference scheme) to infer one step from another by listing it beside the conclusion derived from it.

The item least clear in the list of allowed things is "steps inferred from previous steps," since we haven't said what constitutes a valid deduction form. But surely we know one legal inference: if steps n and $n + 1$ are justified, then so is step $n + 2$:

$n.$ P
$n + 1.$ $P \Rightarrow Q$
$n + 2.$ Q steps n, $n + 1$, M.P.

If you were allowed to know only one legal pattern of deduction, this would be the one; it is called *modus ponens*, hence the "M.P."[3] Another reasonably common one (called *modus tollens*) is as follows:

$n.$ $P \Rightarrow Q$
$n + 1.$ $\neg Q$
$n + 2.$ $\neg P$ steps n, $n + 1$, M.T.

This really may be reduced to *modus ponens* by citing a tautology, and we'll see an example of a proof in the process:

$n.$ $P \Rightarrow Q$ *Hypothesis*
$n + 1.$ $\neg Q$ *Hypothesis*
$n + 2.$ $(P \Rightarrow Q) \Leftrightarrow (\neg Q \Rightarrow \neg P)$ *Tautology*
$n + 3.$ $((P \Rightarrow Q) \Rightarrow (\neg Q \Rightarrow \neg P))$ and
         $((\neg Q \Rightarrow \neg P) \Rightarrow (P \Rightarrow Q))$ Def. $\Leftrightarrow$
$n + 4.$ $(P \Rightarrow Q) \Rightarrow (\neg Q \Rightarrow \neg P)$ and
$n + 5.$ $\neg Q \Rightarrow \neg P$ $n + 1$, $n + 4$, M.P.
$n + 6.$ $\neg P$ steps n, $n + 1$, M.P.

Notice that in line $n + 4$ we used an deduction form to deduce 'A' from 'A and B.' Also we used $\Leftrightarrow$, which is defined by '$P \Leftrightarrow Q$' means '$P \Rightarrow Q$ and $Q \Rightarrow P$.'

We will turn soon to another common form of proof, but we need some notation first. It is common to have to prove results of the form '$P \Rightarrow Q$,' and these may be intermediate results in a long proof. A useful form is as follows (note that this is the *direct* approach to the proof as in Section 2.3, and also that this is a form for the *proof* (that is, a proof form), not the *use* (that is, not an deduction form for the use), of an implication):

    $\cdots$

---

[3]Note that this is the *use* of implication, not the proof of an implication.

$$\begin{array}{cl}
n. & P \quad \text{Assumption} \\
& \cdots \\
m. & Q
\end{array}$$
$m + 1.$ $P \Rightarrow Q$ steps n, m, direct proof

Presumably step m 'Q' rests in some more or less complex way on the assumption n of 'P,' the original hypotheses of the problem, and so on. We enclose steps n through m in a bracket because their later use is risky; step $n + 1$, for example, may depend upon step n. If we use it later we must reassume 'P,' but that may not be obvious. It is helpful to view steps n through m as a subproof of '$P \Rightarrow Q$,' with the bracket a visual reminder that these steps are unavailable for later use.

With this notation in hand, we may discuss the correct form for an argument by cases; this is just the formalization of what was discussed as "common sense" in Section 2.4.1. For example, suppose we desire to prove '$(P$ or $Q) \Rightarrow R$' (a natural-language example might be, "if it rains or snows tomorrow I'll carry my umbrella"). Consider one more time how you would prove this; natural-language arguments provide a reliable guide.

3.3:

Isn't it clear that the form of the proof will be

$$\begin{array}{cl}
j. & P \quad \text{Assumption} \\
& \cdots \\
k & R
\end{array}$$
$k + 1$ $P \Rightarrow R$ j, k, M.P.
$$\begin{array}{cl}
m. & Q \quad \text{Assumption} \\
& \cdots \\
n & R
\end{array}$$
$n + 1$ $Q \Rightarrow R$ m, n, M.P.
$n + 2$ $(P$ or $Q) \Rightarrow R$ $k + 1$, $n + 1$, Pf. by cases

(Of course we proved each of the subimplications directly. What would it look like if the first were by, say, contradiction?) Consider how this would be useful for arguments involving a union of two sets; elements of that union will surely be in one set *or* the other.

3.4:

3.2.1   Exercises

3.5: Give the formal versions of the two other proof forms for '$P \Rightarrow Q$' discussed informally in Section 2.3.

3.6: We stated before that '$P \Rightarrow Q$' is true exactly when '$\neg Q \Rightarrow \neg P$' is true (on this we based the proof form for "proof by contraposition.") Show, using truth tables, that this was correct and the two are equivalent.

3.7: It is useful to be able to exchange the negations of some compound statements for equivalent compound statements. Check via truth tables that '$\neg(P \text{ or } Q)$' is equivalent to '$(\neg P)$ and $(\neg Q)$,' '$\neg(P \text{ and } Q)$' is equivalent to '$(\neg P)$ or $(\neg Q)$,' and that '$\neg(P \Rightarrow Q)$' is equivalent to 'P and $\neg Q$.' Check also that '$P \Rightarrow Q$' is equivalent to '$\neg P$ or Q'. Having checked these, remember them!

3.8: Make a list of deduction forms whose formal versions we have accumulated so far. Pad it out with ones that seem as if they must be right, even though we haven't used them.

The above exercises should make it clear that this section hasn't really been anything new. *IF this were all there is to formal mathematical language*, we could have done things informally as in the previous chapter, because your instincts based on ordinary experience are all correct. But we haven't yet done quantifiers; what we have really covered is "quantifier free logic," whose name is really the predicate calculus. What we've really done here is build a formal structure onto which we can add "quantified logic" (the propositional calculus) in an organized way.

Important! Read Me!

It may help, for what's coming next, to note that for each connective ('and,' 'implies,' ...) we have both an deduction form to *use*, in proofs, statements with that form, and a proof form (or forms) we might use *to prove* such a statement. For an easy example, if we are given the statement 'P and Q,' it is no surprise that we can use this in a proof either to get to 'Q' or to get to 'P,' if we want to. Also, if we are asked to *prove* a statement whose form is 'P and Q,' we expect to first get 'P' (somehow) and then 'Q' (somehow) and our proof form then says we have proved what we want. For each new thing quantifiers let us build, we will have exactly the same pair of forms, one to use, one to prove.

3.3   Quantifiers

Hang on. Here we go.

3.3.1 Statement Forms

Statements such as "the set $(0, 1)$ is open" and "the square function is continuous" will be common in your study of mathematics. These are statements and could be squeezed into what we've mentioned so far, but it would be exceedingly cumbersome to use S for "the square function is continuous," C for "the cube function is continuous," and so on. Much more handy is the use of variables and *statement forms*. For example, let f be a variable understood to range over the set of real-valued functions, and let $C(f)$ be the statement form "f is continuous." This is clearly a handy notation for many assertions about the continuity of numerical functions.

The object $C(f)$ is a *statement form* and not a statement because $C(f)$ is neither true nor false. One way to phrase this is to say that "f" in "$C(f)$" acts as a pronoun, so $C(f)$ might be translated as "it is continuous." One way to get statements from statement forms is as follows: if we let "s" stand for the square function, $C(s)$ is a statement (it happens to be a true one, but what's important is that it is a real statement). If we let "t" stand for the function defined by

$$t(x) = \begin{cases} 1, & x > 0, \\ 0, & x = 0, \\ -1, & x < 0. \end{cases}$$

$C(t)$ is also a statement.[4] It is natural to say that s <u>satisfies</u> <u>the</u> <u>condition</u> C (or <u>has the property</u> C) if $C(s)$ is true.

Pause for Breath

The word "condition" might ring a bell, since we used it informally earlier. Review the beginning of Section 1.3, and see that we finally have the formalization of that idea. Also, you should look at Exercise 1.12, in which we discussed building sets by a "condition." We can now say precisely what we fudged a little there: one way to build a set, via this <u>set</u> <u>builder</u> <u>notation</u>, is to form $\{x : C(x)\}$ where C is a condition, that is, statement form. (Some retreat from this is needed to keep me from getting letter bombs from the logicians. This way to build sets, as stated, is too freewheeling. There are subtle logical problems that require subtle solutions; you'll never run into them, I'll bet. But if your curiosity can't be contained, go read about Russell's Paradox in Lab III, Section 4.3.) Take a little time to get used to this precise formulation of "condition."

---

[4]Objects such as s and t are sometimes called *constants*. This does not mean that they are the constant function. The idea is that f varies over the class of functions (that is, is a variable whose allowed values come from the class of functions), and s and t and *sin* do not so vary, but stand for specific elements of that class.

3.9:

End Pause

Statement forms may be more complex. Exercise your ingenuity by constructing a statement form appropriate for discussing the continuity of a numerical function at a single point.

3.10:

It is a small step to conditions such as $I(f, x_1, x_2)$, where this is notation for the statement form "$f(x_1) > f(x_2)$." The objects the variables range over have been left implicit, as happens frequently.

Finally, it is worth noting that you have seen lots of statement forms before, although you didn't think of them that way. Every equation (e.g., $x^2 + 2x + 1 = 0$) is a statement form; insert any appropriate value for the variable, and you get a statement either true or false. We usually think of an equation as something to be solved, but it is really a statement form. Indeed, the "solution" of an equation is often the replacement of the equation by some other equation that is equivalent to the original (i.e., the values that make one true make the other true, and *vice versa*), but for which it is simpler to read off the values making it true. For example, the statement form above is equivalent to $(x + 1)^2 = 0$, from which it is easier to read off that $x = -1$ is the only value of the variable making the form true. Much of algebra is a list of rules showing what you may do to a statement form that is an equation to transform it to an equivalent one in this way (such as multiply both sides by a nonzero constant). A word to the wise is sufficient, but you might think about inequalities.

3.3.2 Exercises

Formulate appropriate notation to express the following assertions efficiently; express them.

3.11: The graph is connected. How is this different from "the complete graph on four vertices is connected"? Express both.

3.12: The group G is Abelian.

3.13: The relation R is symmetric and reflexive but not transitive.

3.14: If the square function is differentiable at the point 2, then it is continuous at 2.

3.15: If the square function is continuous at the point 2, then it is differentiable at 2.

3.16: The square function is both differentiable and continuous at 2.

3.17: If a relation R is an injective function then R^{-1} is an injective function. Please make "injective" and "function" separate conditions on the basic object a relation.

3.3.3 Quantified Statement Forms

We are still missing one piece of logic necessary for mathematics, namely quantifiers. Memories of calculus may have reminded you that the (true) statement in Exercise 3.14 above is but a tiny bit of a more general fact: for all functions f, if f is differentiable at 2, then f is continuous at 2. Using $D(f, x)$ and $C(f, x)$ for the obvious statement forms, we symbolize this by

$$\forall f(D(f, 2) \Rightarrow C(f, 2)).$$

Note that this *is* a *statement* and not just a statement form. We thus have a way to generate statements from statement forms using the <u>universal quantifier</u> "∀" (read "for all" or " for every" or "for each" or "for any," whichever sounds best).

The other quantifier needed is the <u>existential quantifier</u> "∃" (read "there exists" or "there is"). Recall that $I(f, x_1, x_2)$ was used to denote "$f(x_1) > f(x_2)$." Then

$$\forall f(I(f, 3, 4))$$

is clearly false. But there is a (that is, there *exists* a) function f for which $f(3) > f(4)$, so

$$\exists f(I(f, 3, 4))$$

is true. One place to see how this quantifier might be needed is the Mean Value Theorem, where the point "c" from an earlier discussion calls for it.

Expressions are read from left to right, and in the presence of quantification the order may matter. Consider the expressions

$$\forall x \exists y (y > x)$$

and

$$\exists y \forall x (y > x)$$

with understood domain the real numbers. The first is true, while the second is false. Translations of these into English

3.18:

show that informally the first says that for any number there is a larger number, while the second says that there is a (certain, fixed) number larger than any other. We will discuss these things further when we cover proofs, but in the first expression you may vary y to accommodate various x, while in the second you must get a single y and try to use it for all x.

Finally, some everyday examples should convince you that the negation of quantifiers yields reasonable equivalent expressions:

$$\neg(\forall x(P(x))) \quad \text{is equivalent to} \quad \exists x(\neg P(x)),$$

and

$$\neg\exists x(P(x)) \quad \text{is equivalent to} \quad \forall x(\neg P(x)).$$

Construct some English sentences to convince yourself of this.

3.19:

The negation of long strings of quantifiers is tedious but not difficult if you just work your way in one level at a time:

$$\neg(\forall x(\exists y(\forall z(\forall w(\exists v(\text{ stuff })))))) $$

is equivalent to

$$\exists x(\neg(\exists y(\forall z(\forall w(\exists v(\text{ stuff })))))),$$

which is equivalent to

$$\exists x(\forall y(\neg(\forall z(\forall w(\exists v(\text{ stuff })))))),$$

which is equivalent to

3.20:

What follows is a very small collection of exercises concerning the mechanics of quantifiers. If this is less than you need, do something active about it (consult a text on introductory logic, make up some for yourself, exchange some constructed problems with your classmates).

3.3.4 Exercises

Formulate the following expressions in the language of logical symbols.

3.21: There exists a point c such that $c^2 = 17$.

3.22: There exists a point c such that $f'(c) = 0$.

3.23: For every f, $f(2)$ is an element of $(0, 1)$ and $f'(2) \neq 0$.

3.24: There exists c in $(0, 1)$ such that $f'(c) = 0$.

3.25: There exists c in $(0, 1)$ such that for all x, $f(c) > f(x)$. (This is beginning to get hard.)

3.26: Every group has an identity element.

3.27: Every graph has two vertices with the same degree. (Recall that the degree of a vertex is the number of edges attached to the vertex.)

3.28: There exists a group of order three. Use $G(x)$ for the condition "x is a group" and $O3(x)$ for the condition of having order three (whatever that means).

3.29: Translate the following definitions into appropriate notation.
a) A function f is continuous if it is continuous at each point in its domain.
b) A function f is continuous on the set S if it is continuous at each point of S.

3.30: Negate the following.
a) The sum of the degrees of the vertices of any graph is even.
b) Every basis for $\mathbf{R}^3$ has 3 elements.
c) There exists a matrix with no inverse.

3.31: Negate the following.
a) $\forall x(x \in A \text{ or } x \in B)$.
b) $\exists x(x \notin A \text{ and } x \notin B)$.
c) $\exists y(y \in C \text{ or } y \in D)$.
d) $\forall x(x \in A \Rightarrow x \in B)$.
e) $\forall x(x \in A \Rightarrow x \notin B)$.
f) $\forall y(y \in B \Rightarrow \exists x(x \in A \text{ and } y = f(x)))$.
g) $\forall x_1, x_2(f(x_1) = f(x_2) \Rightarrow x_1 = x_2)$.
h) $(\forall x_1, x_2(g(f(x_1)) = g(f(x_2)) \Rightarrow x_1 = x_2)) \Rightarrow (\forall z_1, z_2(f(z_1) = f(z_2) \Rightarrow z_1 = z_2))$.

3.32: Negate the following statement form:

$$\forall \epsilon(\epsilon > 0 \Rightarrow (\exists \delta(\delta > 0 \text{ and } (\forall x(|x - a| < \delta \Rightarrow |f(x) - f(a)| < \epsilon)))).$$

The results of Exercise 3.7 may be useful.

3.3.5   Theorem Statements

We may take the first step in showing the usefulness of mathematical language in your life by laying out clearly the structure of some theorems, without consideration of their meaning. We use as always the Mean Value Theorem as our first example. To do anything with it, you have to have the statement in front of you. Write it down.

3.33:

The place to start the dissection is to figure out the overall form of the thing. What is it, at a symbolic level? 'P and Q'? 'R or S'? '$\forall v(P(v) \Rightarrow (R(v)$ or $S(v)))$'?

3.34:

It is impossible to continue the discussion without an agreement that the Mean Value Theorem is basically a implication: " *if* (stuff) *then* (other stuff)." As we discussed informally before in Chapter 1 there is the condition of the hypothesis, the condition of the conclusion, and the guarantee given by the theorem.[5] The conclusion is a fairly simple existence claim: there is a number c with two properties, namely that it is in a certain interval and that it satisfies a certain equation in f, f', a, and b. As part of the hypotheses we are told that f is continuous on $[a, b]$ and differentiable on (a, b). (There are also some implicit hypotheses: a and b are clearly intended to be real numbers with $a < b$, f is a real-valued function, and so on.)

Somewhat more sensitivity to language and custom or familiarity with theorems might tell you we have cheated a little in the above description. This is really a theorem about any function f and pair of numbers a and b that satisfy the hypothesis. Thus there are really several universal quantifiers implicit in the statement of the theorem: $\forall f$, $\forall a$, $\forall b$. Note that these quantifiers have to be there. A theorem must be a statement (in fact, a true one!) and without the quantifiers all we have is a statement form.

We need some notation for the various conditions (that is, statement forms): $C(f, S)$ seems reasonable for "f continuous on the set S," and there's a similar notation for differentiability. We might use $L(x, y, z)$ for "x is in the interval (y, z)" and $E(x, y, z, g)$ for

$$g'(z) = \frac{g(y) - g(x)}{y - x}.$$

With these in hand, give the statement of the theorem in symbolic form.

---

[5]The theorem really says that a certain implication is true. Consideration of the truth table for implication shows that, given that the implication is true, we do indeed have what was described as a guarantee before, namely that if the hypothesis holds then the conclusion must also.

3.35:

WARNING: we omitted a needed notation and you may have omitted a hypothesis. Have you ensured $b > a$?

3.36:

This problem is a hard one, since the Mean Value Theorem is complex. If the going is still tough, realize that the form of the MVT is, like that of many theorems, a universally quantified implication. The hypothesis has three terms, and the conclusion two; the conclusion is existentially quantified. It may help to do things first in English before grappling with $L(x, y, z)$ and similar messes.

3.37:

You can always write the hypothesis in the form $H(f, a, b)$ and the conclusion in the form $\exists z (E(a, b, f, z))$, set up the theorem using these building blocks, and then work on each of them separately.

3.38:

This may seem like a lot of work for one theorem, but the breakdown of the MVT into pieces, and the understanding of the structure of the whole, is really useful. We have an implication with hypotheses, conclusion, and quantifications neatly displayed. We are now in a position to examine the relationship between the hypotheses. We can, as in Chapter 1, use this dissection to guide our creation of examples. We can take a hypothesis off in a corner and play with it separately or recall what it means in isolation.[6]

3.3.6 Exercises

Write the following in symbolic form after developing appropriate notation.

3.39: If a set S satisfying $S \subseteq R$ is open, then the set $R - S$ is closed.

3.40: If a group has prime order, then it is simple.

---

[6]Rarely will you read a theorem for the first time with all its components fresh and clear in your mind. We are after the ability to make the theorem "fall apart" so you can deal with trouble spots individually.

3.41: If the sequence $(x_n)_{n=1}^\infty$ converges to L, then it is Cauchy.

3.42: If the sequence $(x_n)_{n=1}^\infty$ is Cauchy, then it converges to some limit.

3.43: If f is continuous on $[a, b]$, $f(a) > 0$, and $f(b) < 0$, there exists c in (a, b) such that $f(c) = 0$.

3.44: Suppose f is a function continuous on $[a, b]$. Then f has a maximum point; that is, there is some c in $[a, b]$ such that $f(c) \geq f(x)$ for all x in $[a, b]$.

3.45: If $\{v_1, \ldots, v_n\}$ is a linearly dependent set of vectors, then for any w, $\{v_1, \ldots, v_n, w\}$ is linearly dependent.

3.46: Every tree with more than one vertex has at least two end vertices. [It may help to note that a "tree" is a special kind of graph, one with the condition of "treeness," whatever that may be. Recall also that the vertex set of a graph G is denoted $V(G)$.]

3.47: Every nontrivial graph has at least two vertices that are not cut vertices. (A nontrivial graph is one with more than one vertex.)

3.48: First we need a definition: a <u>perfect</u> <u>pairing</u> of a graph G is a collection C of ordered pairs of $V(G)$ such that each vertex occurs in exactly one of the pairs and for each pair (v_1, v_2) in C the edge (v_1, v_2) is in $E(G)$. (You should start by expressing this in symbolic language.) Here's the theorem: every graph on $2n$ vertices such that each vertex has degree n or greater has a perfect pairing.

It should be clear after these exercises that the most formal logical language can impede understanding. The phrase "f is continuous on $[a, b]$" is much clearer than "$C(f, [a, b])$" with its accompanying definition, and nobody really wants to say "$L(x, y, z)$" when you could say "x is in the interval (y, z)," let alone "$x \in (y, z)$." Many times the equation, inequality, set containment, or whatever is good enough all by itself. For other things (like continuity, which is none of the above), in real life we use an informal mixture of English and logical symbols. (The accepted mixture is a matter of accepted custom, and its rules are to be learned only by experience. There are some purists who desire either one or the other, and who have lost the fight at least in common practice.) The formal version is much easier to learn and will be useful again when we consider legal steps in proofs.

3.3.7 Pause: Meaning, a Plea, and Practice

We will turn shortly to how quantified statements are used in proofs and how one proves them. But before we do, we'll slip in some propaganda in the guise of practice with quantifiers. On our list of admissible steps in a proof from Section 3.2 you may have viewed "definitions" as a comparatively friendly and familiar item. We call to your attention the observed fact that students underuse drastically this resource in proofs. (The plea part of the title of this section should now be clear.) In sifting through the hypotheses of a theorem one almost invariably arrives at a statement or statement form that is not compound (not 'P and Q,' 'P or Q,' ...) and thus has no obvious subdivision. For example, what can be done with "f is continuous on $[a, b]$" or "G is a group" or "G is a tree" or "$(f_i)_{i=1}^{\infty}$ converges uniformly to f," say?

The thing to realize is that each of these has a *definition*. Each of these *means* something.[7] You may exchange for the hypothesis what it means, so "f is continuous on $[a, b]$" means '$\forall x(x \in [a, b] \Rightarrow f$ is continuous at $x)$,' and so on. These equivalent statements or statement forms give you something new to work with. If faced with the word "tergiversation" in a sentence you may well have to look it up and replace it with what it means (who wouldn't?) and frequently the same thing is useful in a proof. To understand a proof you have to be aware this is what's going on; to do a proof (soon, soon) you need to do it at the right time.

If you did the exercise in the last chapter on the definition of "continuous at a point" you are well prepared to apply this exchange to that definition. First of all, you probably know the definition; write it down.

3.49:

Second, if you are faced with the problem "Prove: f continuous at a implies $a \in domain(f)$" you are prepared to replace f continuous at a with the equivalent but more useful "f has the properties

1. ... ,

2. ... ,   and

3. ... ."

one of which turns out to be exactly what you need.[8] Done. This exchange of the shorthand "f continuous at a" for what it means gives you things to work with (although in this example not much more work is needed).

For another example, let's return to the Mean Value Theorem. Earlier in this chapter we had concluded (with f, a, and b quantified and in place) that the structure of the theorem was '$(C(f, [a, b])$ and $D(f, (a, b))$ and $a < b) \Rightarrow \exists c(L(c, a, b)$ and $E(c, a, b, f))$.' When we used the MVT in the

---

[7]Whether what it means is right on the tip of your tongue isn't the point. You might have to look up or recall something, but there is a definition lurking.

[8]If you understand the definition of $domain(f)$. Here we do it again.

Chapter 1 as a testing ground for the process of constructing examples, we used a different version. The passage from one version to another is really an exercise in meaning (we emphasize that this exercise is really only possible now that the structure of the theorem is laid out).

We may exchange "f continuous on the set $[a, b]$" for '$\forall x(x \in [a, b] \Rightarrow f$ is continuous at x)' and "f differentiable on the set (a, b)" for '$\forall x(x \in (a, b) \Rightarrow f$ is differentiable at x).' Observe that the x of the world fall into two categories, those in both $[a, b]$ and (a, b), and those in (a, b) but not $[a, b]$.[9] Something is a bit curious: for the $x = a$ and $x = b$, we have only one piece of information about the behavior of f. For all of the x's in (a, b) and hence in both sets, we appear to have two pieces of information. What are they?

3.50:

Some memory of a fact from calculus ought to intervene to tell you that if f is differentiable at x it is automatically ... what?

3.51:

Therefore the continuity information given on the interval (a, b) is redundant, and we may assemble an equivalent pair of hypotheses as

1. $\forall x(x \in (a, b) \Rightarrow f$ differentiable at x), and

2. f is continuous at a and b.

This pair (exchanging 1 for the shorthand of the definition) is what we used in the last chapter.

Such insertion of the meaning of conditions (that is, definitions) is a powerful tool, to be neglected at your peril. For a final example, we may turn to a mathematician's favorite, the works of Lewis Carroll.[10] In *Through the Looking Glass, and What Alice Found There* ([3]) we are presented with the word "slithy," and are stuck because we don't know what it means. When informed (by Humpty Dumpty) that it means "lithe and slimy," we may proceed. Many of the objects of mathematical discourse are such "portmanteau" words, and must be understood in the same way.

[9]All right, there are those other x in a third category, namely in neither $[a, b]$ nor (a, b). For such an x, the hypothesis of each of the implications constituting the definition is false, so the implication itself will be true whether its conclusion is true or false. About such x we therefore can't get any information about f's behavior at x.

[10]"Lewis Carroll" was a practicing mathematician. Among his works is *Symbolic Logic* written under his real name, Charles Lutwidge Dodgson. If you need some amusing practice in symbolic logic, browse through his works.

3.3.8 Matters of Proof: Quantifiers

You are probably waiting for the other shoe to drop. You know (mathematical) life isn't the laying out of theorems in a neat logical form, but the proving of them. The more formally you lay them out, the more intimidating they look, and the writing of formal proofs looks out of this world. We start by using the structure already set up; we discussed in Section 3.2 what a proof was, and all we have to do is modify our definition to deal with quantifiers. Recall that we said that a proof was a list of assumptions, steps inferred from previous steps, tautologies, axioms, definitions, and previously proved theorems, culminating in the conclusion we want to prove. To this list we merely add statement forms, the special kind of statement that comes from quantifying a statement form, deduction forms for using them in proofs, and proof forms for proving them.

We surely need such rules, since you can't do mathematics without them. Even to say a function f is continuous on a set S is really to invoke a universal quantifier on the elements of S. Similarly, to prove some statement about all continuous functions is to prove a universally quantified statement. The Mean Value Theorem has both universal and existential quantifiers.

It is perhaps surprising, but should be encouraging, that most parts of many proofs are done at a level not including quantifiers. An informal example from your past experience may show this: consider the proof that every triangle with two congruent angles has also two congruent sides. Prove this theorem. (Please — we need this example to work with.)

3.52:

The usual approach is to draw a triangle containing two congruent angles and proceed. Note that you by no means considered *all* triangles with two congruent angles (how many pictures did you draw, anyway?) but a "generic" one. You have always understood the proof for this "generic" representative of the class of triangles with two congruent angles was enough to prove the result. (Of course, you aren't allowed to behave as if the congruent angles each had measure 30°, since that member of the class of triangles with two congruent angles is no longer generic.) Observe that you dealt with an implicit universal quantifier, not by examining all triangles, but by proving the result for a "fixed but arbitrary" such triangle; the real work was done with that triangle. We will formalize this powerful argument scheme (proof form) in a bit. It's perfectly suited for *proving* theorems with universal quantifiers.[11]

---

[11] If this sort of universal quantification were all, an informal discussion of quantifiers would be enough. This "pick a generic one and prove" is pretty harmless, but things get worse.

You probably also *used* in the proof a universally quantified theorem about similar triangles: if two triangles have ... then the two are similar. Note that the general idea is that you may use this theorem on any appropriate pair of triangles of your choice, in particular the pair in your theorem. Here is an inference pattern (deduction form) clearly suited to the *use* of hypotheses or theorems with universal quantifiers.

A little thought makes it clear that we need (at least) four procedures for dealing with quantifiers. We need two patterns of argument for *proving* results, one to be a form for proving results with universal quantifiers, one to be a form for proving results with existential quantifiers; that is, we need proof forms for the two types of quantified results. We also need two procedures for *using* universally quantified and existentially quantified results (that might, perhaps, appear in the hypothesis of what we are trying to prove); that is, we need deduction forms for the two types of quantified statements. We'll begin with the ways to *use* the two flavors of results.

Universally Quantified Results: Use

Suppose you have as a hypothesis '$\forall x(P(x))$.' What may you deduce — what can you get out of this? The answer is of course '$P(y)$,' where y is any object in the appropriate domain of the variable:

1. $\forall x(P(x))$ Hypothesis
2. $P(y)$ U.I.

This argument scheme is <u>Universal Instantiation</u>, which we have already abbreviated to U.I. We will come later to matters of how to use the scheme effectively.[12] The scheme itself is straightforward: if you know that all female duck-billed platypi lay eggs, you may deduce that some particular duck-billed platypus named Raquel lays eggs.

Cueing of U.I.

Since U.I. is used to deduce something about a specific element from the universal, its use is a one-step event, so there are no beginning and ending cues. One might say something like "Therefore, applying the universal to y, ..." or "applying the hypothesis to y ..."; often the clue is that you are *applying to* some specific element.

Existentially Quantified Results: Use

Suppose you have as a hypothesis '$\exists x(P(x))$.' What may be deduced? It is helpful to think of the scheme as a "naming" process. There is *something*

---

[12]This is a very powerful tool, since you get to use it on *any* appropriate y. The skill in using it is picking a y, or sometimes the only y, for which knowing $P(y)$ is useful.

out there with property P, so call it (something):

1. $\exists x(P(x))$ Hypothesis
2. $P(x_*)$ E.I.

Here E.I. stands for <u>Existential</u> <u>Instantiation</u>. We adopt here and will use for a while the convention of subscripting by $*$ the names introduced by an application of E.I.

This scheme requires more care than that for U.I. One (informal) way to say it is that the name introduced must be chosen so as not to conflict with other symbols already occurring in the proof. Suppose we have a proof containing the steps shown:

$n.$ $G(f)$
$n + 1.$ $G(f) \Rightarrow H(f)$
$n + 2.$ $H(f)$ $n, n + 1$, M.P.
        $\ldots$
$m.$ $\exists g(Q(g))$ Hypothesis

We are certainly allowed to give a name to an object having property Q, but in this proof, we'd better not choose the name f. To use f would be to assume that the object f, which we already have and are stuck with, is one of the objects (perhaps the only object) with property Q.[13] The good notational habit of not using a symbol in two different ways is usually safeguard enough. Put simply, when you produce an object with some property by using E.I., give it a new name.

Cueing of E.I.

E.I. allows you to know that a certain object exists, and you almost always want to give a name to that object early in the proof so you can use it; again this is a one-step event. One might say "Let y_* be the (whatever) whose existence we are given," or simply "Denote by y_* the (whatever)." Observe that "let," "denote," and "set" now have a use besides setting up notation, which is to use E.I. and set the notation at the same time. WARNING: in ordinary English, "there exists" is the same as "there is," and the same is true here. But "there is" is a less clear cue to an existence statement, and more easily overlooked, so watch carefully for it.

We turn next to schemes for proving statements with quantifiers.

---

[13]You have a lottery ticket, and you know that somewhere there exists a winning lottery ticket. Is it safe to assume that yours is the one with the special property? No, alas.

Existentially Quantified Results: Proof

How might one conclude '$\exists x(P(x))$' (that is, prove there is something with property P)? A legal way, and in practice the only way, is to establish '$P(y)$' for some (specific) y. If I manage to exhibit a green pig with wings, you must admit that one exists:

$$\vdots$$

$n.$	$P(y)$	
$n+1.$	$\exists x(P(x))$	E.G.

Here *E.G.* stands for <u>Existential</u> <u>Generalization</u>. There is almost never any other way to prove a conclusion of this form; we will turn later to where to look for some such y. Since examples are always useful, we point out that the Mean Value Theorem, the Intermediate Value Theorem, and the Maximum Theorem (see 1.3.1, 1.4.1, and 1.8.3, respectively, although all are from first-term calculus) are all existence results.

Cueing of E.G.

Usually E.G. is cued in two places, beginning and end: it is kind to indicate to the reader that you plan to show something exists by producing it, and it's nice to remark at the end that you succeeded. Perhaps "We will construct the (something) needed ..." and "and so y is the (something) desired" are reasonable beginning and ending cues.

Universally Quantified Results: Proof

This proof form is the most subtle and probably most important. Let us start with a rule too freewheeling but with the right intuitive flavor. Frequently, for example, we wish to prove something of the form '$\forall x(H(x) \Rightarrow J(x))$.' (Think, for example, of the result you proved above about triangles.) Consider the following argument, where we have again adopted a subscripting convention:

$n.$	x_0	Fixed but arbitrary
$n+1.$	$H(x_0)$	Assumption
	$\ldots$	
$m.$	$J(x_0)$	
$m+1.$	$H(x_0) \Rightarrow J(x_0)$	$n+1, m$, direct proof
$m+2.$	$\forall x(H(x) \Rightarrow J(x))$	$n, m+1$, U.G.

Here *U.G.* stands for <u>Universal</u> <u>Generalization</u>.

Informally, we have chosen x_0 a "fixed but arbitrary" object in the understood domain, and since we wish to prove $m+1$ (an implication) we assume $H(x_0)$ as usual. We then prove $J(x_0)$ as usual by some arguments free of the universal quantification involving x. We then argue that since x_0 was a generic object with property H, and for it we could deduce property J, and hence $H(x_0) \Rightarrow J(x_0)$, we may deduce line $m + 2$. Recall the proof you gave a while back about some triangle; this is exactly the formalization of that argument. Note that we have adopted a subscripting convention for Universal Generalization in which we subscript by "0" the arbitrary item(s) we are using for this proof form.

The language surrounding the use of U.G. in mathematics written in paragraph form is, frankly, terrible. First, we said above that x_0 was "fixed but arbitrary." This use of "arbitrary" is unusual, because it does not mean that *you* may choose it arbitrarily (that is, at your decision) but that it is an arbitrary choice you are handed. Second, proofs written in English may signal the intent to use this argument by some phrase such as "Let f be a continuous function" or "Let $\epsilon_0 > 0$ be given." "Let" really ought to be "assume"; "given" really means "given to you and you are stuck with it." This language is confusing when you are just starting out, but everybody uses it and therefore so must you.[14]

Occasionally one faces a proof for which no hypothesis seems appropriate; the result isn't a universally quantified implication but simply a universally quantified statement form $(\forall x(P(x)))$. For example, it is true that for any pair of sets A and B, $A \cup B = B \cup A$. The common beginning of a proof written in paragraph form would be "Let A_0 and B_0 be arbitrary sets." Our convention that the domains of variables remain implicit really gets in our way here; the theorem statement is really

$$\forall A, B(A, B \text{ sets } \Rightarrow (A \cup B = B \cup A)).$$

This made explicit, we may proceed with the beginning given.

This inference scheme U.G. is great; too bad it's not right (as presented). Consider the following argument, where $C(f, 2)$ and $D(f, 2)$ are, as before, continuity and differentiability at the point 2:

1. $\exists f(C(f, 2) \Rightarrow D(f, 2))$ Hypothesis
2. $C(f_*, 2) \Rightarrow D(f_*, 2)$ E.I.
3. $\forall f(C(f, 2) \Rightarrow D(f, 2))$ U.G.

The hypothesis in step 1 is true.[15] Step 3 is false, since one can surely construct a function continuous at 2 but not differentiable there. The form of the argument must therefore be flawed, since a valid argument cannot lead from a true hypothesis to a false conclusion. Rats.

---

[14]Sorry about that.

[15]For the square function, for example, '$D(f, 2)$' is true, so the implication '$C(f, 2) \Rightarrow D(f, 2)$' is true (think of the truth table for '$\Rightarrow$') and so there does indeed exist an f for which the implication is true.

Arguments from the particular to the general are delicate. It just can't be correct to deduce from "there exists something with property P" the conclusion "everything has property P." Intuitively one should use U.G. on a statement form with the symbol to be quantified "fixed but arbitrary," and the result of E.I. is not arbitrary in the sense of "generic." Our subscripting conventions help here, since anything with the subscript $*$ is inappropriate for the use of U.G., while anything with subscript 0 is at least a candidate.

Unfortunately, we still aren't out of the woods. Consider the following argument, with the variables understood as real numbers:

1. $\forall x(\exists y(x < y))$ Hypothesis
2. $\exists y(x_0 < y)$ U.I.
3. $x_0 < y_*$ E.I.
4. $\forall x(x < y_*)$ U.G.

What a wonderful y_*! It seems to be larger than all real numbers.[16] Step 1 is correct, yet we are in trouble again. Intuitively, the x_0 introduced in step 2 was "generic," but the y_* introduced in step 3 is not generic since it clearly depends on x_0. In step 2 x_0 was completely unrestricted: we could insert any real number for x_0. In step 3, since y_* has been chosen, some choices of x_0 are clearly ruled out. In some sense, x_0 is no longer generic.[17]

This one is rather distressing, for the subscripting convention doesn't really help here. The good thing is that most of the time the "freewheeling" version of U.G. really works, at least if used in accordance with the subscripting conventions. Very rarely in real (mathematical) life do you find yourself constructing an argument of the type to get you in the second kind of trouble. (On most of the few occasions that you do, the result you prove will be so outrageous that you will know something went wrong.) Most of the time the freewheeling version of U.G. will get you through just fine. If you wish to understand more deeply, you need to take a course in symbolic logic (preferably mathematical logic) and come to understand things called "bound" and "free" variables.

Cueing of U.G.

The language isn't good, as we have said, but at least it is standard. Very often one starts with "Let y_0 be arbitrary" or "For any y_0, ..." and ends with "Since y_0 was arbitrary, we have the result in general" or "Since the result holds for each y_0, we have the result in general" or something of the sort. Occasionally a short argument might start "For each y_0, ..." The key word is often *arbitrary*, although sometimes students are taught to use the word *generic* instead. (Advantage: the word is more descriptive. Disad-

---

[16]Ouch.

[17]The rule for avoiding this difficulty is the hardest to state.

vantage: nobody else uses it.) WARNING: this proof form is so standard that frequently cues are omitted altogether, and one simply starts with a y_0 and shows what is needed for it, leaving the reader to recognize the form and fill in the quantification and deduction scheme U.G.

We now have four argument forms, two for using quantified results and two for proving them. While there are technical restrictions, most of them aren't too bad. Even better, it will turn out that there is rarely any choice about what to do. We'll talk in the next section about how often there is only one door around to be opened, and how to spot what that door might be.

3.3.9   Exercises

The goal in these exercises is to return to some previous problems (the exercise numbers in parentheses) and spot all the quantification arguments you may have missed the first time out. So on top of previously identified proof structures (proof by contradiction, for example) you need to find all the E.I., U.I., E.G., and U.G. arguments around.

3.53: (2.17). Suppose that f and g are functions for which the composition $h = g \circ f$ is defined. If h is injective, then f is injective.

Proof. Suppose that f is not injective. By definition, then, there are distinct points x_1 and x_2 in the domain of f so that $f(x_1) = f(x_2)$. Then clearly $h(x_1) = (g \circ f)(x_1) = g(f(x_1)) = g(f(x_2)) = h(x_2)$, and therefore h is not injective.

3.54: (2.18). The identity element in a group is unique.

Proof. Let G be a group and suppose e and f act as identity elements. Then $e = e * f$ since f is an identity, and $e * f = f$ since e is an identity. Combining these equations, $e = f$ and so these elements are the same.

3.55: (2.19). The identity element in a group is unique.

Proof. Let G be a group and suppose e and f are distinct elements that act as identities. Then $e = e * f$ since f is an identity, and $e * f = f$ since e is an identity. Combining these equations, $e = f$, a contradiction.

3.56: (2.20). Let $\{A_\alpha\}$ be a collection of connected sets with nonempty intersection. Then $\cap_\alpha A_\alpha$ is connected.

Proof. Let p be a point in the intersection, and suppose that $\cap_\alpha A_\alpha$ is not connected. Then there exists a disconnection C, D as usual; in particular, neither C nor D is empty and $\cap_\alpha A_\alpha = C \cup D$. Without loss of generality, we may assume that $p \in C$. Now for each α, $p \in A_\alpha$, and therefore $C \cap A_\alpha$ is not empty. Therefore, since A_α is connected, A_α is contained entirely in C. It follows that D is empty, which is a contradiction.

3.57: (2.21). Let S be a set of real numbers bounded above, and denote by $-S$ the set $-S = \{-x : x \in S\}$. Then $-S$ is bounded below.

Proof. Let b be an upper bound for S. For any y in $-S$, there is an x in S such that $y = -x$. Further, $x \leq b$ since b is an upper bound for S, and so of course $-b \leq -x$. Thus $-b \leq y$, and it follows that $-b$ is a lower bound for $-S$.

3.58: (2.22). Let $\{v_j\}_{j=1}^n$ be a linearly dependent set of vectors in a vector space V (with $n \geq 2$). Then there is a subset of the v_j with $n - 1$ elements with the same span as the original $\{v_j\}_{j=1}^n$.

Proof. We construct the required subset. Since the set $\{v_j\}_{j=1}^n$ is linearly dependent, there exist scalars $\{\lambda_j\}_{j=1}^n$, not all zero, so that $\lambda_1 v_1 + \lambda_2 v_2 + \ldots + \lambda_n v_n = 0$. Without loss of generality, we may assume that $\lambda_1 \neq 0$. We then claim that the set $\{v_j\}_{j=2}^n$ has the required properties, and clearly the number of elements is correct. To show that its span is the same as that of the full set, suppose v is in the span of $\{v_j\}_{j=1}^n$, so there exist scalars $\{\alpha_j\}_{j=1}^n$ such that $v = \alpha_1 v_1 + \alpha_2 v_2 + \ldots + \alpha_n v_n$. From the dependence equation, we have $v_1 = -\lambda_2/\lambda_1 v_2 - \ldots - \lambda_n/\lambda_1 v_n$, and it is easy to deduce from the last two equations that v is in the span of the $\{v_j\}_{j=2}^n$. We have therefore constructed a set with the desired properties.

3.59: (2.23). Cographs of isomorphic graphs are isomorphic.

Proof. Let G and H be graphs, and h an isomorphism between them. By definition, we must show that there exists an isomorphism between G^c and H^c; we will show that h is such an isomorphism. Surely h is a one-to-one function from the vertex set of G^c onto that of H^c, since the vertex sets of G and G^c, and H and H^c, respectively, are the same. We next must show that for any two vertices v_1 and v_2 of G^c, the edge v_1–v_2 is in G^c if and only if the edge $h(v_1)$–$h(v_2)$ is in H^c. Suppose v_1–v_2 is in G^c; then by definition, v_1–v_2 is not in G. Since h is an isomorphism of G and H, $h(v_1)$–$h(v_2)$ is not in H. By definition, $h(v_1)$–$h(v_2)$ is in H^c, as desired. The other implication is similar, and we are done.

3.60: (2.24). If a is an element of a group such that $a * a = a$, then a is the identity element.

Proof. Let e denote the identity of the group. There exists b such that $b * a = e$. Observe $b * (a * a) = e$ from our assumption on a. Also, $(b * a) * a = a$. Therefore, using associativity, we have $a = e$ as desired.

3.61: (2.25). Suppose that $a_1 + \ldots + a_9 = 90$, with the a_i nonnegative integers. Then there exist three of the a_i whose sum is greater than or equal to 30.

Proof. WLOG, we may suppose the a_i are in decreasing order. If $a_1 + a_2 + a_3 \geq 30$, we are done. If not, then clearly $a_4 + a_5 + a_6 < 30$ and $a_7 + a_8 + a_9 < 30$, a contradiction.

3.62: (2.26). Suppose f and g are functions for which the composition $g \circ f$ makes sense. If f and g are injective, then $h = g \circ f$ is injective.

Proof. We use the definition of injective directly. Suppose that x_1 and x_2 are elements such that $h(x_1) = h(x_2)$. Then by the definition of h, $g(f(x_1)) = g(f(x_2))$. Since g is injective, we may deduce $f(x_1) = f(x_2)$, and since f is injective, we may deduce $x_1 = x_2$, as desired.

3.63: (2.27). The intersection of an arbitrary nonempty family of subgroups of a group is again a subgroup.

Proof. We use the theorem stating that a nonempty subset S of G is a subgroup if and only if for every a and b in S, $a * b^{-1}$ is in S. Let $\mathcal{S}$ denote the family of subgroups and T denote the intersection. To show T nonempty, note that since the identity element of G is in S for each $S \in \mathcal{S}$ it is surely in T. Suppose a, b are in T. Then a is in S for each $S \in \mathcal{S}$ and similarly for b. So for each S, $a * b^{-1} \in S$ citing the theorem. Then $a * b^{-1} \in T$ from the definition of intersection, and we are done via the theorem.

3.64: (2.28). Let G be a graph, and define a relation R on the vertex set $V(G)$ by $(a, b) \in R$ if and only if there is a walk from a to b. Then R is an equivalence relation.

Proof. We verify the three conditions for an equivalence relation. First, there is the trivial walk from any vertex to itself, so $(a, a) \in R$ for each a. Second, suppose there is a walk from a to b. By reversing the order of the list of vertices, we may obviously produce a walk from b to a, which is symmetry. Finally, suppose (a, b) and (b, c) are in R. By simply concatenating the lists of vertices for the two walks, we may obviously produce a walk from a to c, which is transitivity. Thus we are done.

3.65: (2.29). If G is a group and a is any fixed element of G, we define $T_a : G \to G$ to be the function of right translation by a, so $T_a(x) = x * a$ for all $x \in G$. Prove that T_a is injective.

Proof. We use the definition of injective, so suppose $T_a(x) = T_a(y)$. Then $x * a = y * a$. Multiplying on the right of each side of this equation by a^{-1} it is easy to deduce $x = y$ as required.

3.66: (2.31). For any sets A, B, and C, $(A \cap C) \cup (B \cap C) \subseteq (A \cup B) \cap C$.

Proof. Let x be an arbitrary element of $(A \cap C) \cup (B \cap C)$. If $x \in A \cap C$, we have $x \in A$ and $x \in C$. Then surely $x \in A \cup B$ and $x \in C$, and therefore $x \in (A \cup B) \cap C$. The other case is similar.

3.67: (2.32). Suppose d is a metric on a space M, and we define d_b by $d_b(x, y) = \min(d(x, y), 1)$. Then d_b is a metric on M.

Proof. We must check the three conditions for a metric. Since $d_b(x, y)$ is the minimum of nonnegative quantities it is nonnegative, and it can only equal zero if $d(x, y) = 0$, which implies $x = y$ as needed. Also, $d_b(x, y) =$

$\min(d(x,y),1) = \min(d(y,x),1) = d_b(y,x)$ since d is a metric. For the third, we must show that for any x, y, and z,

$$d_b(x,y) \le d_b(x,z) + d_b(z,y).$$

If either $d_b(x,z)$ or $d_b(z,y)$ is one, the inequality is obvious, so we turn to the case $d_b(x,z) < 1$ and $d_b(z,y) < 1$. Now $d(x,y) \le d(x,z) + d(z,y)$ since d is a metric. Then from this and our assumption, we have $d(x,y) \le d_b(x,z) + d_b(z,y)$. Combining this with $d_b(x,y) \le d(x,y)$, we have the result desired in this case as well.

3.68: (2.33). For any $n \ge 1$, the sequence n, n, $n-1$, $n-1$, $\ldots$, 2, 2, 1, 1 is graphic. (See Exercise 1.92 for the needed definition.)

Proof. It is clear that the sequence 1, 1 is graphic: just connect two vertices by an edge. Suppose that for all j less than or equal to n, the sequence j, j, $j-1$, $j-1$, $\ldots$, 2, 2, 1, 1 is graphic; we must show that $n+1$, $n+1$, n, n, $n-1$, $n-1$, $\ldots$, 2, 2, 1, 1 is graphic. Since $n-1$, $n-1$, $\ldots$, 2, 2, 1, 1 is graphic, there exists some graph G for which it is the degree sequence. We shall construct the desired graph H by adjoining some vertices and edges to G. Collect the vertices of G into two subsets A and B, each of which has one vertex of degree $n-1$, one of degree $n-2$, and so on down to 1. We add 4 vertices to G to produce H. Add vertices x_1 and y_1. Add two more vertices X_1 and Y_1; connect by edges X_1 to every vertex in A and to x_1, Y_1 to every vertex in B and to y_1; and put an edge between X_1 and Y_1. It is easy to check that the resulting graph has the required degree sequence.

3.69: (2.34). Definition: for x a real number, we define $|x|$ by $|x| = x$ if $x \ge 0$ and $|x| = -x$ if $x < 0$. Prove that for any x, $|-x| = |x|$.

Proof. If $x = 0$ the result is trivial. If $x > 0$, then $-x < 0$, so by definition $|-x| = -(-x) = x = |x|$. If $x < 0$, then $-x > 0$, and $|-x| = -x = |x|$ again by definition.

3.70: (2.35). Suppose L is a collection of n straight lines in the plane, no two parallel and with $n \ge 2$, and such that no more than two lines meet at any point. Then there are $n(n-1)/2$ points of intersection of the lines in L.

Proof. It is trivial to verify the formula for $n = 2$. Suppose the formula holds for some $k > 2$. To show it holds for $k+1$, remove one line from the collection. It is trivial to verify that what remains is a collection of k lines, no two parallel, and such that no more than two lines meet at any point. By hypothesis, then, there are $k(k-1)/2$ points of intersection of this smaller collection of lines. Now reintroduce the deleted line. Since it is parallel to none of the k others, it has k points of intersection with them, and none of these has already been counted since in that case there would be a point on three lines, disallowed. So there are $k(k-1)/2 + k$ points of intersection, and a little algebra completes the result.

3.71: (2.36). Suppose M is a metric space with metric ρ. Suppose A is a dense subset of B and B is a dense subset of C. Show that A is a dense subset of C.

Proof. We use the definition: recall that X is a dense subset of Y if for every $y \in Y$ and every $\epsilon > 0$ there exists $x \in X$ so that $\rho(x, y) < \epsilon$. Therefore we must show that for every point c of C and every $\epsilon > 0$ there exists a point a of A such that $\rho(a, c) < \epsilon$. So let c and $\epsilon > 0$ be arbitrary. Since B is dense in C, using the definition applied to c and $\epsilon/2$ there exists $b \in B$ such that $\rho(b, c) < \epsilon/2$. And since A is dense in B, using the definition applied to b and $\epsilon/2$ there exists a in A so that $\rho(a, b) < \epsilon/2$. It is then easy by the triangle inequality to deduce that $\rho(a, c) < \epsilon$, so a is the point desired. Since c was arbitrary in C and $\epsilon > 0$ was arbitrary, we are done.

3.72: (2.37). Prove that in any group G, $(a * b)^{-1} = b^{-1} * a^{-1}$.

Proof. Observe that $(a * b) * (b^{-1} * a^{-1}) = e$ by an easy application of associativity. Therefore $b^{-1} * a^{-1}$ is an inverse of $a * b$. Since inverses are unique, it must be "the" inverse $(a * b)^{-1}$.

3.4   Finding Proofs from Structure

A Dose of Propaganda

Let's suppose for the moment that you think the formal language of symbolic logic isn't too bad.[18] The other shoe still hasn't dropped. As the title of this section indicates, you will have to do more than state theorems or note that proofs use the rules of inference correctly. The heart of mathematics is finding the darn things in the first place. The more formal we get, the worse that looks.

This section is to start convincing you that the structure of a theorem statement frequently gives at least an outline of the proof. Professors know this; most times when a professor sees a proof as easy or obvious, and a student has no idea how to proceed, the professor is seeing the clues in the structure that almost "force" the proof and the student isn't. Sometimes you can do a proof completely by following the clues. Sometimes you will at least get through some routine things and save your brain for where an idea really is needed. If sensitive to these clues, you will spend comparatively little time just floundering around.

Let's consider some examples. If a result you need to prove has the form

$$\forall f(f \text{ a function } \Rightarrow \text{ stuff})$$

then almost always you will need to pick f an "arbitrary" function and show

---

[18]For example, perhaps you are better able to follow what is going on in a proof involving quantifiers.

that (stuff) holds. If you want to prove something about all functions, you may in theory either

1. examine all functions, one by one (out of the question!), or

2. use the standard U.G. argument form, proving the result for a "fixed but arbitrary" representative.

The description above is, of course, the freewheeling version; we anticipate lines that, in a formal proof, would be:

n.	Assume f_0 is a function	Assumption
	$\ldots$	
m.	stuff about f_0	
$m + 1$.	$\forall f(\text{stuff about } f)$	$n, m,$ U.G.

We'll write proofs in this section less formally than the line-by-line version. (Although it is good practice to write, or at least rewrite, one in the formal way once in a while.) But whatever the format, a proof of the result about functions above will always be some form of U.G.

It is not just that U.G. works; it is that any other approach is likely to be off in left field somewhere. *Perhaps* there is another theorem somewhere that says all your objects have the desired property, but then you are out to prove an extremely boring theorem. Almost always, you ignore the clue given by the form of the result (in this case, universally quantified) at your peril. Another example where you need U.G. is proofs involving the definition of continuity ("For every $\epsilon > 0 \ldots$"). If you begin a problem of the type "Prove using the definition that f is continuous, where $f(x) = \ldots$" any way *other* than "Assume $\epsilon > 0$ is arbitrary $\ldots$," you will be headed for frustration (bets on this are gladly accepted by the author). The form of the result gives the good clue.

Consider another example in which the structure of the conclusion dictates the form of the proof. Suppose the conclusion you want is of the form 'P or Q.' It is just about inconceivable that the proof won't have an argument that is, or could be presented as, two cases, one of which leads to 'P' and one of which leads to 'Q.'[19] If you ignore the clue, you may not look for cases, which is fighting with one brain tied behind your back. Incidentally, if the argument does take this form, there may well be an "or" in the hypothesis, each branch of which is an assumption for one case. Note that here is a proof structure indicated even without considering quantifiers.

There's yet another example in which the form of the conclusion dictates the form of the proof: existence conclusions [for example, "there exists c in (a, b) such that $f'(c) = (f(b) - f(a))/(b - a)$"]. You may twist and turn

---

[19]Think of natural-language examples — what about a "proof" that tomorrow it will either rain or snow? Aren't there cases having to do with the temperature?

all you like, but your efforts must be directed toward producing such a c. Very occasionally, one can do this by producing some c explicitly, as we did in Chapter 1 with $c = 7/2$ for the Mean Value Theorem. Much more frequently, you need to find some hypothesis or theorem guaranteeing the existence of some other point d with some properties (that is, you need another existence result). Perhaps d will serve as the c you need. Perhaps $d/2$ will work as c. In more complicated cases, there may be two relevant existence results, one yielding d and the other e, with the c you want some combination of the two. The point is again that there is really only one path to the existence result you want, and that clue is too useful to ignore.

The structure of the hypothesis, as well as that of the conclusion, can dictate the form of the proof. (The author has a preference, certainly personal, for looking first at the conclusion.) If the hypothesis is 'P or Q' then it is likely that the argument divides into two cases: Case 1 ('P' holds) and Case 2 ('Q' holds). If one of your hypotheses is that "$\forall \epsilon > 0$ (something)" then you are undoubtedly going to have to apply this to some (perhaps cleverly chosen) $\epsilon_0 > 0$. The hypothesis structure clues are more guides to the form of the proof.

One of the great frustrations for professors is the student who works agonizingly long hours doing things that must ultimately be characterized as irrelevant to the problems.[20] Sensitivity to the structural clues in the hypothesis and conclusion of results to prove will vastly decrease this wasted time. Frequently there are few if any reasonable choices for the structure of the proof, which is good and not bad unless you like exploring the forest without a compass.

3.4.1 Finding Proofs

Let's try to do (that is, discover) a proof *mechanically* by following the clues of structure, particularly quantifier structure, and inserting meaning (that is, definitions) when necessary.

$$\text{Prove } A \cap (B \cup C) \subseteq (A \cap B) \cup (A \cap C).$$

Work: we realize first that there are assumed universal quantifiers, since this is likely to be about general sets. We may therefore write the theorem formally as

$$\forall A \forall B \forall C (A \cap (B \cup C)) \subseteq (A \cap B) \cup (A \cap C))$$

where we assume that A, B, and C are sets. Very well, the proof must have the first step

Pf. Let A_0, B_0, and C_0 be arbitrary sets.

---

[20]There are persistent rumors in the teaching profession that this is frustrating for the student as well.

Let's throw in the desired conclusion and get a proof outline that looks like this:

> Pf. Let A_0, B_0, and C_0 be arbitrary sets.
> ...
> So $A_0 \cap (B_0 \cup C_0) \subseteq (A_0 \cap B_0) \cup (A_0 \cap C_0)$ as desired.

Since A_0, B_0, and C_0 were arbitrary sets,
the result holds in general.   ∎

We have here an outline of a proof founded on U.G., and our first line signals the intent to use such an argument. Notice that we need work only inside the box, since the universal quantifiers have been taken care of by the form of the argument.

We don't seem to have gotten very far. **Look at the conclusion.** It can't be logically subdivided, so we'll need to insert its meaning. What does

$$A_0 \cap (B_0 \cup C_0) \subseteq (A_0 \cap B_0) \cup (A_0 \cap C_0)$$

mean? We need the definition of "$\subseteq$"; inserted in our situation, this is

$$\forall x(x \in A_0 \cap (B_0 \cup C_0) \Rightarrow x \in (A_0 \cap B_0) \cup (A_0 \cap C_0)).$$

In order to conclude "So $A_0 \cap (B_0 \cup C_0) \subseteq (A_0 \cap B_0) \cup (A_0 \cap C_0)$ as desired" the line just above must be the definition above. So we have the slightly improved proof outline:

> Pf. Let A_0, B_0, and C_0 be arbitrary sets.
> ...
> Thus, $\forall x(x \in A_0 \cap (B_0 \cup C_0) \Rightarrow x \in (A_0 \cap B_0) \cup (A_0 \cap C_0))$.

So $A_0 \cap (B_0 \cup C_0) \subseteq (A_0 \cap B_0) \cup (A_0 \cap C_0)$ as desired.
Since A_0, B_0, and C_0 were arbitrary sets,
the result holds in general.   ∎

We see immediately that the conclusion we hope to get ("Thus, ...") is universal, and the required method of proof is forced:

Pf. Let A_0, B_0, and C_0 be arbitrary sets.

> Let x_0 be arbitrary.
> ...
> So $x_0 \in A_0 \cap (B_0 \cup C_0) \Rightarrow x_0 \in (A_0 \cap B_0) \cup (A_0 \cap C_0)$.

Thus, $\forall x(x \in A_0 \cap (B_0 \cup C_0) \Rightarrow x \in (A_0 \cap B_0) \cup (A_0 \cap C_0))$.
So $A_0 \cap (B_0 \cup C_0) \subseteq (A_0 \cap B_0) \cup (A_0 \cap C_0)$ as desired.
Since A_0, B_0, and C_0 were arbitrary sets,
the result holds in general.   ∎

Note that we need work only within the box, since the rest is taken care of by the form of the proof (another U.G.).

Aha! (Well ... or something.) The form of the conclusion is an implication, and we know how those have to be proved: assume the hypothesis, deduce the conclusion (well ... at least if we are using direct proof[21]). Thus we will use the direct implication proof form. So we arrive at:

Pf. Let A_0, B_0, and C_0 be arbitrary sets.
Let x_0 be arbitrary.

> Suppose $x_0 \in A_0 \cap (B_0 \cup C_0)$.
> ...
> So $x_0 \in (A_0 \cap B_0) \cup (A_0 \cap C_0)$.

So $x_0 \in A_0 \cap (B_0 \cup C_0) \Rightarrow x_0 \in (A_0 \cap B_0) \cup (A_0 \cap C_0)$.
Thus, $\forall x (x \in A_0 \cap (B_0 \cup C_0) \Rightarrow x \in (A_0 \cap B_0) \cup (A_0 \cap C_0))$.
So $A_0 \cap (B_0 \cup C_0) \subseteq (A_0 \cap B_0) \cup (A_0 \cap C_0)$ as desired.
Since A_0, B_0, and C_0 were arbitrary sets,
the result holds in general.   ■

Our work is now confined to the box. The conclusion is logically indivisible (that is, it is not a compound statement or statement form), so let's insert what it means. We need the definition of "$\cup$" applied to our situation. Inserting this before our conclusion, we arrive at:

Pf. Let A_0, B_0, and C_0 be arbitrary sets.
Let x_0 be arbitrary.

> Suppose $x_0 \in A_0 \cap (B_0 \cup C_0)$.
> ...
> Hence $x_0 \in (A_0 \cap B_0)$ or $x_0 \in (A_0 \cap C_0)$.

So $x_0 \in (A_0 \cap B_0) \cup (A_0 \cap C_0)$.
So $x_0 \in A_0 \cap (B_0 \cup C_0) \Rightarrow x_0 \in (A_0 \cap B_0) \cup (A_0 \cap C_0)$.
Thus, $\forall x (x \in A_0 \cap (B_0 \cup C_0) \Rightarrow x \in (A_0 \cap B_0) \cup (A_0 \cap C_0))$.
So $A_0 \cap (B_0 \cup C_0) \subseteq (A_0 \cap B_0) \cup (A_0 \cap C_0)$ as desired.
Since A_0, B_0, and C_0 were arbitrary sets,
the result holds in general.   ■

Again the work remaining is confined to the box. Note also that we have made substantial progress. **Look at the conclusion**: since it is an 'or,' it is likely that the argument divides into cases, one with the conclusion $x_0 \in (A_0 \cap B_0)$ and the other with the conclusion $x_0 \in (A_0 \cap C_0)$. So the proof must look like this:

Pf. Let A_0, B_0, and C_0 be arbitrary sets.
Let x_0 be arbitrary.

---

[21] Note that this was the first place we had a real choice.

Suppose $x_0 \in A_0 \cap (B_0 \cup C_0)$.

. . .

Case I.

. . .

So $x_0 \in (A_0 \cap B_0)$.
Case II.

. . .

So $x_0 \in (A_0 \cap C_0)$.
Hence $x_0 \in (A_0 \cap B_0)$ or $x_0 \in (A_0 \cap C_0)$.

So $x_0 \in (A_0 \cap B_0) \cup (A_0 \cap C_0)$.
So $x_0 \in A_0 \cap (B_0 \cup C_0) \Rightarrow x_0 \in (A_0 \cap B_0) \cup (A_0 \cap C_0)$.
Thus, $\forall x(x \in A_0 \cap (B_0 \cup C_0) \Rightarrow x \in (A_0 \cap B_0) \cup (A_0 \cap C_0))$.
So $A_0 \cap (B_0 \cup C_0) \subseteq (A_0 \cap B_0) \cup (A_0 \cap C_0)$ as desired.
Since A_0, B_0, and C_0 were arbitrary sets,
the result holds in general.   ■

We may start with the hypothesis to see how to get to these cases. Since $x_0 \in A_0 \cap (B_0 \cup C_0)$ is logically indivisible, we should insert what it means. This is, by definition, "$x_0 \in A_0$ and $x_0 \in B_0 \cup C_0$." That $x_0 \in A_0$ is clearly relevant, and also clearly "atomic" — there is nothing more to be done with it than to use it.[22] What about "$x_0 \in B_0 \cup C_0$"? By definition (again) this means "$x_0 \in B_0$ or $x_0 \in C_0$."

Before we insert these, we need to recall one more fact. What we have is: "$x_0 \in A_0$ and ($x_0 \in B_0$ or $x_0 \in C_0$)." The general form of this is 'P and (Q or R).' From Section 3.2 of this chapter you may remember that this is equivalent to '(P and Q) or (P and R),' and also that it is sometimes necessary to exchange some statement for an equivalent one. Take a moment and figure out, for our particular P, Q, and R, what the equivalent statement is (time for you to do some work).

3.73:

Now we have the following update of the proof, where the 'or' in the statement you found gives the key to the cases:

Pf. Let A_0, B_0, and C_0 be arbitrary sets.
Let x_0 be arbitrary.
Suppose $x_0 \in A_0 \cap (B_0 \cup C_0)$.
Then $x_0 \in A_0$ and $x_0 \in (B_0 \cup C_0)$.
Thus $x_0 \in A_0$ and ($x_0 \in B_0$ or $x_0 \in C_0$).

---

[22]Speaking very carefully, "$\in$" is one of the *undefined notions* from which all of mathematics is derived logically. Less formally, that a certain thing is an element of a certain set is about as simple as it gets. Use it.

So $(x_0 \in A_0$ and $x_0 \in B_0)$ or $(x_0 \in A_0$ and $x_0 \in C_0)$.
Case I. $x_0 \in A_0$ and $x_0 \in B_0$

. . .

So $x_0 \in (A_0 \cap B_0)$.
Case II. $x_0 \in A_0$ and $x_0 \in C_0$

. . .

So $x_0 \in (A_0 \cap C_0)$.
Hence $x_0 \in (A_0 \cap B_0)$ or $x_0 \in (A_0 \cap C_0)$.

So $x_0 \in (A_0 \cap B_0) \cup (A_0 \cap C_0)$.
So $x_0 \in A_0 \cap (B_0 \cup C_0) \Rightarrow x_0 \in (A_0 \cap B_0) \cup (A_0 \cap C_0)$.
Thus, $\forall x(x \in A_0 \cap (B_0 \cup C_0) \Rightarrow x \in (A_0 \cap B_0) \cup (A_0 \cap C_0))$.
So $A_0 \cap (B_0 \cup C_0) \subseteq (A_0 \cap B_0) \cup (A_0 \cap C_0)$ as desired.
Since A_0, B_0, and C_0 were arbitrary sets,
the result holds in general. ∎

To the extent that there is anything left to fill in, fill it in.

The point is that we proceeded through the proof quite mechanically. We could nibble away at the proof, discovering a layer at a time, each layer forced or at least strongly indicated by the preceding and following steps. Structural clues helped us through; here is a way of attacking proofs somewhere between instant inspiration and complete inability to get started. A good deal goes into this method, but it is a great many small steps rather than a try for one big one. There are two ways to fight monsters: one is to rush the monster, sword swinging, in a single do or die attempt. The other is to try to maneuver around, slicing off a toe here and a finger there. One has a substantially greater life expectancy associated with it.[23]

Aside

It should now be clear why one might present the proof, in good faith, by including only what will fill the box in the paragraph beginning "Aha!". That would rest on an assumption that the audience would recognize all the universal quantifier mechanics omitted and could supply it on request. See Section 2.2 for a similar example.

End Aside

Exercise

Try the above process out on the following.

3.74: Prove $(A \cap B) \cup (A \cap C) \subseteq A \cap (B \cup C)$.

Let's try another example of this. Consider the following theorem:

---

[23]There is one point we have glossed over. In the proof above, when a definition was needed we inserted it smoothly and went on. In real life, *you* will have to realize you need one, remember or go find it, and insert it. This isn't completely trivial, but it is a *skill* you can *learn*, not a "have it or don't" *talent*.

Theorem *Let f be a function. For any A and B subsets of domain(f) we have*

$$f(A \cup B) \subseteq f(A) \cup f(B).$$

If you need some relevant definitions, they are in a previous chapter. Take some time and figure out the structure of the theorem. Then set up the first (and last) steps of the proof.

3.75:

We hope you realized that there are implicit universal quantifiers on f and on the sets A and B. Other than that, the theorem is a straightforward implication:

$$\forall f \forall A \forall B (A, B \subseteq domain(f) \Rightarrow f(A \cup B) \subseteq f(A) \cup f(B)).$$

Therefore you arrive, since we know the proof form for universal quantifiers and how to prove an implication (we'll try the direct proof form), at the following:

Pf. Let f be an arbitrary function and A and B be arbitrary sets.

> Assume $A, B \subseteq domain(f)$.
> . . .
> Thus $f(A \cup B) \subseteq f(A) \cup f(B)$.

Since A and B were arbitrary sets, the result holds for all A and B subsets of $domain(f)$.
Since f was arbitrary, the result holds in general.   ■

Note that we have not followed the convention of subscripting f, A, and B by 0. This was done primarily for ease of reading, but also the author is getting tired of doing it. Nonetheless we will use U.G. on f, A, and B and trust that a mental notation of that intent may replace the subscripting. From now on we use the convention only as it seems helpful.

What next?

3.76:

We hope you looked at the conclusion and realized it has logically indivisible structure, so it is time to insert some meaning.[24] Time for the definition of "$\subseteq$" again.

---

[24]If you looked at the hypothesis, you weren't wrong, but there isn't a whole lot to be gotten from it in this particular proof.

3.77:

Having inserted that, you should realize that we are faced with a rather familiar form of the conclusion and set yourself up to deal with it.

3.78:

We hope you found something rather like this:

> Pf. Let f be an arbitrary function and A and B be arbitrary sets.
> Assume $A, B \subseteq domain(f)$.

> Assume $y_0 \in f(A \cup B)$.
> ...
> Thus $y_0 \in f(A) \cup f(B)$.

> So $y_0 \in f(A \cup B) \Rightarrow y_0 \in f(A) \cup f(B)$.
> Then $\forall y(y \in f(A \cup B) \Rightarrow y \in f(A) \cup f(B))$.
> Thus $f(A \cup B) \subseteq f(A) \cup f(B)$.
> Since A and B were arbitrary sets, the result holds for all A and B subsets of $domain(f)$.
> Since f was arbitrary, the result holds in general.   ■

Let's pause for a moment and look at the above partial proof. It looks rather impressive and unlikely to be something that you would at first feel that you could do. The point is that this step-by-step approach has cut it into bite-sized pieces. We never have to do very much at a time. (Indeed, a cynical person might say that we've gotten this far without any honest labor at all.) OK, back to work.

How are we able to conclude something is in a union? Gee ... we might need the definition. Upon its insertion, what is the form of the argument you might expect?

3.79:

Perhaps you got something like the following:

> Pf. Let f be an arbitrary function and A and B be arbitrary sets.
> Assume $A, B \subseteq domain(f)$.

Assume $y_0 \in f(A \cup B)$.

...

Case I.

...

Thus $y_0 \in f(A)$.
Case II.

...

Thus $y_0 \in f(B)$.
Thus $y_0 \in f(A)$ or $y_0 \in f(B)$.

Thus $y_0 \in f(A) \cup f(B)$.
So $y_0 \in f(A \cup B) \Rightarrow y_0 \in f(A) \cup f(B)$.
Then $\forall y(y \in f(A \cup B) \Rightarrow y \in f(A) \cup f(B))$.
Thus $f(A \cup B) \subseteq f(A) \cup f(B)$.
Since A and B were arbitrary sets, the result holds for all A
and B subsets of $domain(f)$.
Since f was arbitrary, the result holds in general.   ■

How do we connect to the cases? We still haven't arrived at the level of
what $f(A)$ or $y_0 \in f(A)$ mean, and we probably should. Might it be time
for a definition? Insert the appropriate definitions of $f(A)$ and apply it to
y_0.

3.80:

Pf. Let f be an arbitrary function and A and B be
arbitrary sets.
Assume $A, B \subseteq domain(f)$.
Assume $y_0 \in f(A \cup B)$.
Then by definition, $\exists x(x \in A \cup B$ and $f(x) = y_0)$.

Let x_* be some such x, so $x_* \in A \cup B$ and $f(x_*) = y_0$.

...

Case I.

...

Therefore $\exists z(z \in A$ and $f(z) = y_0)$.
Thus $y_0 \in f(A)$.
Case II.

...

Therefore $\exists w(w \in B$ and $f(w) = y_0)$.
Thus $y_0 \in f(B)$.
Thus $y_0 \in f(A)$ or $y_0 \in f(B)$.

Thus $y_0 \in f(A) \cup f(B)$.
So $y_0 \in f(A \cup B) \Rightarrow y_0 \in f(A) \cup f(B)$.

Then $\forall y(y \in f(A \cup B) \Rightarrow y \in f(A) \cup f(B))$.
Thus $f(A \cup B) \subseteq f(A) \cup f(B)$.
Since A and B were arbitrary sets, the result holds for all A
and B subsets of $domain(f)$.
Since f was arbitrary, the result holds in general.   ∎

To finish making the connection, we need either some insight (mechanical
is good, but insight is OK!) or some careful following of the needs of an
existence conclusion. We must show in Case I that there exists a z having
two properties, one that $z \in A$, the other that $f(z) = y_0$. As we remarked
in our discussion of proofs of existence results, this is probably going to be
done by the *exhibition* of such a z. That is, to come up with an existentially
quantified conclusion, we have to use the deduction form for existential
statements. Where is such an object to come from? We have so far in the
proof only f, A, B, x_*, and y_0 as fixed objects. It should be clear that f,
A, and B are out of the running as candidates for z. Why?

3.81:

Also, y_0 is not a good candidate, either. Why?

3.82:

We hope you said that f, A, and B were the wrong sort of object (being
functions or sets, not elements of the domain of a function), and that y_0
is in the codomain $codomain(f)$, not the domain $domain(f)$.[25] This puts
the burden on x_*, which at least does half of the job, since $f(x_*) = y_0$. A
sensible person would hope that x_* really is satisfactory, for which it would
have to be in A. Perhaps sometimes it is: why don't we make this Case I?

> Case I. $x_* \in A$
> Then $x_* \in A$ and $f(x_*) = y_0$.
> Therefore $\exists z(z \in A$ and $f(z) = y_0)$.
> Thus $y_0 \in f(A)$.

Case II is equally clear, where we assume instead that $x_* \in B$, thus making it a prime candidate for the use of E.G. to get $\exists w(w \in B$ and $f(w) =
y_0)$. Are there any more cases? Must it be true that either $x_* \in A$ or
$x_* \in B$? Decide why or why not and fill in the rest of the proof.

3.83:

---

[25] Alternatively for disposing of y_0, it should seem unlikely that $f(y_0) = y_0$. Of
course this sort of thing happens, but is it *likely*?

As we look back over the proof[26] there are a number of remarks to make. First, the fact that x_* was the useful thing for existence results really isn't a surprise; as we noted when we discussed existence results, often the object resulting from a hypothesis giving an existence *conclusion* is the object you need. Secondly, the proof as finally discovered this way may not be awfully well written, but the rewriting is the easy part.

Further, we used considerable care in notation that may have passed unnoticed but is really important. For example, we took care to use variable names fitting our intuition and habits: y for things in the range of f, x for things in the domain, capital letters for sets, and so on. If you are sloppy in picking names, the mechanics will fight your intuition; being careful may actually force you to think about something important. To choose variable names well is a "proof discovery" version of one of our rules of thumb about writing proofs.

We were also careful to use $\exists z(z \in A$ and $f(z) = y_0)$ instead of $\exists x(x \in A$ and $f(x) = y_0)$ and it is worthwhile to point out why. To use the latter notation carries the psychological if nonmathematical assumption that x_* or at least x-something will turn out to be the z we need. In this case it was, but you will run into proofs in which it is $x_*/2$ or something even worse. Particularly if you drop the subscripting conventions as you become more comfortable with these arguments, it is very hard for nonmathematical reasons to write:

   n. $x/2$ satisfies $P(x/2)$
   m. $\exists x(P(x))$.

One often writes in a proof in paragraph form "So there exists an x, namely ( ), satisfying ..." where ( ) is the specific one you found. To write "there exists x, namely $x/2$, ..." is even worse than doing it in a formal proof. (A discussion of bound and free variables solves the problem completely on a mathematical level and not at all on a psychological level.) So we write $\exists z(z\ldots)$ and if x is a satisfactory z, well and good; if we have to use $x/2$ or $x/2 + y/3$ or something, an intervening step like $z_0 = x/2 + y/3$ is comforting.

The final remark in our *postmortem* of the proof is that you might have felt cheated near the end. The mechanics near the end didn't give a completely clear form for the proof, and we had to do some honest labor. A more positive way to put it is that the mechanics forced us to concentrate exactly at the point where something inventive was needed. (And, really, a good understanding of existence proofs did take us home.) The mechanics saved us a good deal of flailing around that might have prevented us from coming close to the key point. The proof as a whole is a pretty nice object;

---

[26] If you don't do this, but having gotten the proof proceed to shut the book and go watch television, you are missing a small amount of work with a large payoff.

mechanics helped us write it step by step.

It would be good practice to rewrite this in the formal style:

1. Statement 1   Reason 1

$n.$ ....

Some markers for subproofs are appropriate.

3.84:

3.4.2 Exercises

3.85: Prove for any sets A, B, and C that

$$A \cap (B \cup C) \subseteq (A \cap B) \cup (A \cap C).$$

You must certainly use, so may need to find, the definition of set containment.

3.86: Prove for any sets A, B, and C that

$$A \cup (B \cap C) \subseteq (A \cup B) \cap (A \cup C).$$

3.87: Prove that for any sets A and B that

$$(A \cup B)' \subseteq A' \cap B'.$$

It will be necessary to remember or track down the definition of the complement of a set.

3.88: Prove carefully and using the definition of injective that the function $f : \mathbf{R} \to \mathbf{R}$ given by $f(x) = 3x - 7$ is injective.

3.89: Prove carefully and using the definition of surjective that the function $f : \mathbf{R} \to \mathbf{R}$ given by $f(x) = 3x - 7$ is surjective on $\mathbf{R}$.

3.90: Prove carefully and using the definition of injective that if the function $f : \mathbf{R} \to \mathbf{R}$ is injective then the function $g : \mathbf{R} \to \mathbf{R}$ defined by $g(x) = f(x) + 2$ for all x is injective.

3.91: Prove carefully from the definition of surjective that if the function $f : \mathbf{R} \to \mathbf{R}$ is surjective then the function $g : \mathbf{R} \to \mathbf{R}$ defined by $g(x) = f(x) + 2$ for all x is surjective on $\mathbf{R}$.

3.4.3 Digression: Induction Correctly

Back in Section 2.4.3 we gave an intuitive version of proof by induction that may have been what confused you when you first tried them: "to prove something by induction you first prove it for the case $n = 1$ and then, assuming its truth for n, prove it for $n + 1$." The expression many students give for their confusion is "If I am trying to prove it for all n, in the second step where I assume it is true for n, aren't I assuming what I am supposed to prove?" With the symbolic logic in place, and now that you are more comfortable with the deduction forms, this can be put to rest.

Here is the Induction Theorem:

Theorem 3.4.1 *Let $P(n)$ be a condition with n (a positive integer variable) the only free variable.[27] To prove $\forall n \geq 1(P(n))$ it is enough to prove*

1. $P(1)$, and

2. $\forall n \geq 1(P(n) \Rightarrow P(n+1))$.

Without worrying about the proof of this theorem, how do you use it? Well, of course, there is the proof of $P(1)$. Next, you have to prove $\forall n \geq 1(P(n) \Rightarrow P(n + 1))$. How will you do this? By U.G., of course, which involves choosing an n_0 fixed but arbitrary and proving $P(n_0) \Rightarrow P(n_0+1)$. And how do you prove this implication? By assuming $P(n_0)$ and deducing $P(n_0 + 1)$.

Notice that the usual description of induction, in its attempt never to mention quantifiers, gives you only about half of the steps that are really going on. Further, by not making the distinction between n (a variable) and n_0, a single specific fixed but arbitrary value of that variable, it makes induction impossible to think about; the more you think, the worse it gets. Observe that in the full proof above you are by no means assuming $\forall n(P(n))$, which would indeed be cheating. All you are doing is using a proof form that involves the temporary assumption of a single $P(n_0)$ (and that form is for an implication you didn't even know was there before this discussion). By the way, some texts try to describe induction by "to prove for all n, $P(n)$ by induction you first prove it for the case $k = 1$ and then, assuming its truth for k, prove it for $k + 1$." This is an effort to avoid quantifiers while making the confusion above disappear by n versus k. It doesn't really work, but it is an attempt to get at n versus n_0. It's better just to bite the bullet and talk about quantifiers.

You may not believe in induction any more than you used to, since the Induction Theorem itself is one we won't prove. But the hope is that at least what you are supposed to do doesn't seem like complete nonsense.

---

[27]Wait! Don't panic! This is just to appease the logicians; pretend I said "Let $P(n)$ be a statement about n."

Exercises

3.92: Prove, carefully and formally, that for each positive integer n,

$$1 + 2 + \ldots + n = \frac{n(n+1)}{2}.$$

3.93: Prove formally that for each n,

$$\frac{1}{1 \cdot 2} + \frac{1}{2 \cdot 3} + \ldots + \frac{1}{n(n+1)} = \frac{n}{n+1}.$$

3.94: Prove that for each positive integer n,

$$3^{n+3} > (n+3)^3.$$

3.95: Formulate, in formal quantified terms, what we called "weak" induction in Section 2.4.4.

3.96: Define the Lucas sequence by $L(1) = 1$, $L(2) = 3$, and $L(n) = L(n-1) + L(n-2)$ for $n > 2$. Define the Fibonacci sequence as usual by $F(0) = 1$, $F(1) = 1$, and $F(n) = F(n-1) + F(n-2)$ for $n \geq 2$. Prove that $L(n) = F(n) + F(n-2)$ for $n \geq 2$.

3.97: Prove that every integer greater than 1 is a product of primes.

3.4.4 One More Example

We give, on a last theorem, one more example of this general step-by-step approach. WARNING: The tools of structure and meaning (formal language and definitions) will not be enough to get us all the way through the proof.

Theorem *Let X be a set, and A and B subsets of X. If $A \subseteq B$ then $\chi_A \leq \chi_B$.*

We start as usual with the statement of the theorem in formal language:

3.98:

Something should be clear: before we can complete the proof of the theorem, and very possibly before we can get very far, we must understand all of the terms in the theorem's statement. What is χ_A? What does it mean to say $\chi_A \leq \chi_B$, or, in general, $f \leq g$ where f and g are functions?

   Of course, there is a definition lurking in the background, and here it is:

Definition *Suppose f and g are real-valued functions satisfying the equation domain$(f) =$ domain(g). We say $f \leq g$ if*

$$\forall x(x \in \text{domain}(f) \Rightarrow (f(x) \leq g(x))).$$

Intuitively, the value of f at each point is smaller than or equal to the value of g at the same point. You ought to remember from some earlier chapter or other that it is vital to do something to become at home with the definition. Do it, both with general f and g, and limiting yourself to χ_A for various sets A. Shame on you if you don't draw some pictures.

3.99:

If by chance you don't remember the definition of χ_A, you need to look it up.[28] Try Exercise 1.3 in Chapter 1.

3.100:

Before returning to the proof, let's make two points. One is that in upper-level mathematics you will frequently be called upon to view a function as a single object, rather than a collection of values. The above definition is certainly in this spirit. As with the above definition, however, the definition of a property frequently reduces to a statement that the same property holds for all the values of the function. The property in this case is $\leq$-ness of functions, which reduces to $\leq$-ness of the values of the function.

You might also note that the development of the proof would have been much smoother if we had gone through the definition of $f \leq g$ and encouraged you to recall the definition of χ_S in advance. It doesn't always happen that way. In fact, it doesn't even usually happen that way. **It is unrealistic to believe you will always bring to your first attack on a theorem everything you need.** The "interruptions" of going back to understand a definition more clearly or working with some examples are really part of the business. Our attack on this proof, diversions and all, is more realistic than our previous presentations, not less.

Well, where are we? We want to prove

$$\forall X \forall A \forall B ((A, B \subseteq X \text{ and } A \subseteq B) \Rightarrow \chi_A \leq \chi_B).$$

We now know what all this means. It is clear U.G. is going to get a workout. Set up the proof as far as clearing out the quantifiers goes.

3.101:

There is also an implication to deal with. How will you take care of it?

---

[28]Of course, you remembered this general rule about getting started. Indeed, we are employing it to help get started on a theorem, whose statement contained an unclear term. Now we are going to use it on the subproblem of that term.

3.102:

It must be hopeless to proceed without the definition of "$\leq$" for functions applied to χ_A and χ_B, so we insert this. We are then faced with more U.G. Continue the process.

3.103:

We hope you arrived at something rather like the following:

> Assume X, A, and B are sets.
> Suppose $A \subseteq X$ and $B \subseteq X$.
> Assume also that $A \subseteq B$.
> Suppose that $x_0 \in X$. (Note that $X = domain(f)$.)
> . . .
> Thus $\chi_A(x_0) \leq \chi_B(x_0)$.
> So $x_0 \in X \Rightarrow \chi_A(x_0) \leq \chi_B(x_0)$.
> Hence $\forall x(x \in dmn(f) \Rightarrow \chi_A(x_0) \leq \chi_B(x_0)$.
> Therefore $\chi_A \leq \chi_B$.
> Since X, A, and B were arbitrary sets, the result
> holds in general.

We seem to be running out of mechanical things to do. We haven't yet used the fact that $A \subseteq B$, nor have we inserted its meaning. It is tempting to insert it, but it means

$$\forall z(z \in A \Rightarrow z \in B).$$

This merely delays the problem, for we need to find a z to apply this to. What is the only candidate in sight?

3.104:

The problem is that there is no particular reason to believe that $x_0 \in A$. Of course, the implication is still true if $x_0 \notin A$, but it will be true whether $x_0 \in B$ or $x_0 \notin B$, which doesn't seem very useful. We do seem to need some sort of idea here. Spend some time looking for one.

3.105:

Aside from the mathematical ideas you might have found, there is a "process of proving" idea you might have had. What did we do in the last proof when x_0 in a certain set would have been useful? Go find out, and then try it here. What sort of an argument are we building now?

3.106:

What you find is what might be called Case I.[29] Here it is:

Note that $\forall z(z \in A \Rightarrow z \in B)$.
Case I. $x_0 \in A$
Then we have $x_0 \in B$.
So $\chi_A(x_0) = 1$.
Also $\chi_B(x_0) = 1$.
Thus $\chi_A(x_0) \leq \chi_B(x_0)$.

What about another case for $x_0 \notin A$? The examples you constructed should have convinced you that both $x_0 \in B$ and $x_0 \notin B$ might hold. Well, there seem to be two more cases; what are you waiting for? Complete the proof, while you are at it.

3.107:

Looking back over the proof, we feel obligated to say that somebody will point out that in the case $x_0 \notin A$, two "subcases" depending on whether x_0 is in, or not in, B aren't necessary. From $x_0 \notin A$ you may deduce $\chi_A(x_0) = 0$. Since $\chi_B(x_0)$ is either 0 or 1, clearly $\chi_A(x_0) \leq \chi_B(x_0)$. This is neater but perhaps less obvious.

Try out these techniques on the following:

3.4.5 Exercises

3.108: Let X be a fixed set. For $A \subseteq X$, define A' (the *complement* of A), by

$$A' = \{x : x \in X \text{ and } x \notin A\}.$$

Prove that

$$\chi_{A'} = 1 - \chi_A.$$

[HINT: You need to start with some understanding of what "1" means. Since 1 is a number and $\chi_{A'}$ and χ_A are functions, we seem to have an equation with both apples and oranges. In this context the symbol 1 is used to mean the constant function whose value is always 1:

$$1(x) = 1 \quad \text{for all } x.$$

The notation is indeed poor, but it is also common.

---

[29] As a matter of fact, what else could it be called?

Also, it is worth noting that "prime" is used for things other than the complement. For example, you know it is used for derivative (f'). It is also sometimes used to provide another name for a variable, say, x and x'. Sensitivity to context and some common sense are required.]

3.109: Suppose R is a relation on a set S that is symmetric and transitive, and such that for each s in S, s occurs as an element of at least one ordered pair in R. Prove that R is reflexive.

It may be helpful to formulate the definitions of reflexive, symmetric, and transitive (see Exercise 1.24) so as to make the universal quantifiers explicit.

3.110: Suppose R and R_1 are equivalence relations on a set S. Prove that their intersection, $R \cap R_1$, is also an equivalence relation on S.

3.111: Suppose f and g are functions for which the composition $g \circ f$ is defined, and suppose f and g are injective. Prove that $g \circ f$ is injective.

3.112: Suppose f and g are functions for which the composition $g \circ f$ is defined, and suppose that composition is injective. Prove that f is injective.

3.113: Suppose f and g are functions for which the composition $g \circ f$ is defined, and suppose that f and g are surjective on their respective codomains. Prove that $g \circ f$ is surjective.

3.114: Suppose f and g are functions for which the composition $g \circ f$ is defined, and suppose that the composition $g \circ f$ is surjective. Prove that g is surjective.

3.115: Return to Exercises 2.17–2.29 and 2.31–2.37 and take another try at finding all the quantification forms. It may help to rewrite them using the conventions on subscripting; you may find that there was more than you caught the first and second times.

3.5   Summary, Propaganda, and What Next?

What are a few of the things that go into this "mechanical" technique for finding things? First, you have to understand symbolic logic well enough both to see the structure of a theorem and to be sensitive as to how that structure helps determine the proof. Second, you have to know definitions well enough to insert them when meaning is required. (As well, you have to be able to handle the interruption inevitable when you go back and look up a definition for insertion.) A dead end will result if you don't; it isn't shown in these examples, but the same thing will happen if you don't know past theorems well enough to spot when they might be useful. Third, you must be willing to *write things down* even before you are sure they are right or relevant. It is easier not to have to juggle three unwritten things in mind as you try to think; in particular, writing down the conclusion gives you something to aim for. Finally, you have to be committed to a step-by-step and active approach to doing proofs; it is time to give up "inspiration or bust." Part of that commitment is a willingness to make mistakes, recover from them, and in general engage in discovery. Any sort of real discovery is a creative process requiring willingness to work at it. This technique is designed to take that willingness and get good results from it.

As with our discussion of the "examples approach" in Chapter 1, there are some obvious modifications here. If, on a given day, you happen to have some inspiration or insight, go ahead and form the proof about the "right idea."[30] It should be clear that many of these steps can be combined or partially omitted or inserted only in the final write-up, especially as you become more familiar with proofs. Indeed, much of this will be done mechanically and not consciously, leaving you to concentrate on what really needs concentration.

It should be no surprise that exploiting formal structure to find proofs won't do all the proofs in the world. Let's look at a particularly convincing example, which is the Schröder–Bernstein theorem.

Theorem *Let A and B be sets. If there exists an injective function f from A to B, and an injective function g from B to A, then there exists a one-to-one correspondence between A and B.*

Take a shot at proving this.

3.116:

The understanding of formal structure *does* help you know that to show there exists a one-to-one correspondence you must produce (construct) one, leads you to suspect that you must use *f* and *g* somehow, and perhaps helps you understand how you might use the injectivity of *f* and *g* when it comes time to prove that your candidate works. Constructing some examples and drawing some pictures may have helped you understand the problem better too. But these alone won't give you (or at least most of us) enough help to do the proof, because a nontrivial *idea* is required.

In this sense the techniques taught in this book are like techniques in art. Bad techniques hamper you; good techniques let you concentrate your energies on what really matters, in the proof reading or painting analysis, and in the proof discovery or painting creation. But techniques *alone* cannot turn you into a creative prover or a creative artist. That sort of ability is

---

[30]This is all about what to do on the other days. It is sad but true that there are lots of other days.

the result of years of dedicated practice and some wild card called "talent," whatever that is and wherever it comes from.

Most of us are not going to become those incredible intellects who conquer new and lofty heights of previously undiscovered or unproved theorems. Techniques are important for everybody (even them!), but are especially so for the rest of us as we use lesser talents to make our use of mathematics in other areas, or the reasoning skills we've learned in mathematics in other situations, and so on. One way to view the aim of this book is to think of it as attempting to remove flaws in your technique so you may go as far as your talent and desire take you.

But you might still want to understand and prove the Schröder–Bernstein Theorem (or its analog in some other area of mathematics you find more appealing). After you have mastered these techniques of exploiting formality to do what it can, where do you concentrate your energies next? Of course, you take more courses, learn more content, accumulate more experience, see more proofs, and so on. But where do you put your *conscious* effort on learning to *prove better* (as opposed to learning some particular new proof)?

I believe that the next set of mental calisthenics is a conscious effort to "see it at a glance," as Pólya says in *How To Solve It* [5, page *xvii*]. The following proof of the Schröder–Bernstein Theorem gives an example of what is needed (this particular write-up was taken from Ref. [6], where it is a nonexample of how to present proofs so that they may be seen at a glance):

Proof. Define $X_0 = g(B)$, $\varphi = g \circ f$, $Y_0 = A - X_0$ and (recursively) $Y_n = \varphi(Y_{n-1})$ for all $n > 0$. Now let $Y = \cup_{n=0}^{\infty} Y_n$, $X = A - Y$. Note that $Y_0 \subseteq Y$, so $X \subseteq X_0 = g(B)$, hence $g^{-1}(a)$ is defined for $a \in X$ (since g is one-to-one). Note, too, that the definition of X guarantees that $X \cup Y = A$ and $X \cap Y = \emptyset$. Thus the following rule gives a well-defined function $h : A \to B$:

$$(3.1) \qquad h(a) = \begin{cases} f(a), & a \in Y, \\ g^{-1}(a), & a \in X. \end{cases}$$

Let $y \in Y$ be given. Then $y \in Y_n$ for some $n \geq 0$, hence $\varphi(y) \in \varphi(Y_n) = Y_{n+1} \subseteq Y$. Thus $\varphi(Y) \subseteq Y$. Suppose $h(a_1) = h(a_2)$ and distinguish three cases. If $a_1, a_2 \in Y$ then $f(a_1) = f(a_2)$ hence $a_1 = a_2$. If $a_1, a_2 \in X$ then $g^{-1}(a_1) = g^{-1}(a_2)$, hence $a_1 = a_2$. In the remaining case we have, say, $a_1 \in X$ and $a_2 \in Y$. Then $g^{-1}(a_1) = f(a_2)$, hence $a_1 = g(f(a_2)) = \varphi(a_2) \in \varphi(Y) \subseteq Y$. Since the result $a_1 \in Y$ contradicts $X \cap Y = \emptyset$, this case cannot happen. We have seen that in all cases $h(a_1) = h(a_2)$ implies $a_1 = a_2$, that is, h is one-to-one. Let $b \in B$ be given. We put $a = g(b)$ and distinguish two cases. If $a \in X$, then $h(a) = g^{-1}(a) = g^{-1}(g(b)) = b$. If $a \notin X$ then $a \in Y$, hence $a \in Y_n$ for some $n > 0$. ($a \in Y_0$ is impossible since $a \in g(B) = X_0$.) It follows that $a \in \varphi(Y_{n-1})$, hence $a = \varphi(y)$ for some

$y \in Y$. Now $g(h(y)) = g(f(y)) = \varphi(y) = a = g(b)$ and since g is one-to-one, $h(y) = b$. We have found a source for b in all cases, so h is "onto." This completes the proof of the theorem.

Well, in spite of what you may think, the work you have done in this book will allow you to follow the steps in this proof if you are willing to spend the time. But unless you spend a considerable amount of time after that, you will never either find or understand the idea behind all these manipulations. The endeavor to dig out and state simply the crucial idea behind the proof is, I believe, the effort to which you must discipline yourself next. It is hard not to stagger over the finish line of understanding the mechanics of this proof and immediately swear off ever thinking about it again, but you'll never *understand* (as opposed to *follow the steps in*) the proof if you don't. Your goal should be to get to some idea like this (again from Ref. [6]):

> Partition A as $A = X \cup Y$, and take h to be f on X and g^{-1} on Y. By choosing the partition wisely, we hope to make the function h well-defined, one-to-one and "onto".

Diligence in looking for this sort of idea (something like the "top-down" approach in computer programming — see Ref. [6] for a discussion) in proofs you read and identifying the idea in proofs you discover is the next step.

A more traditional answer to "What next?" would surely be "Read *How To Solve It*, by Pólya" [5]. It is beyond doubt that this classic book is full of good habits to acquire, or that it strikes a chord with many mathematicians as a perfect description of what we do, or that teachers of mathematics are filled with hope for its effects on their students. Every mathematician or would-be mathematician ought to read it, and you should. But you should be aware as well that its prescriptions may be too vague and general (however intuitively correct) to follow unaided. For example, in a list of suggestions and in a few pages here and there Pólya discusses making an example. Chapter 1 of this book is the expansion of this into something my students found concrete and detailed enough to use in practice.

There is another issue to be considered for "What next?" as well, and that is how to control all these various techniques you may learn. A beautiful and enlightening discussion of these problems is in Schoenfeld's *Mathematical Problem Solving* [7]; very briefly, you run into the problem of choosing wisely and then continually monitoring your progress with the various techniques and approaches at your disposal. If you've ever (who hasn't) wasted large amounts of time on an exam or in your room following a blind alley to the exclusion of all else, you've experienced one of these problems. Similarly, if you tend to dive into calculations before you know what you are going to do with them, you've experienced another. Someone once said that the three most embarrassing questions to ask of a student working on a mathematics problem are "What are you doing?", "Why are you doing it?", and "What are you going to do with it when you get it?" One change that might improve your problem-solving skills is to make sure you are answering those questions internally all the time.

Schoenfeld's book is fascinating and important reading for any teacher of mathematics (and perhaps any mathematician); it is, however, mostly research oriented and by no means a self-improvement book. One place to find some exercises and a thorough discussion of ways to improve higher-level problem-solving and proof skills is in the material of the McMaster Problem Solving Program, at least some of which was written by D. R. Woods of McMaster University. Many of the exercises are perfect for work with a partner interested in the same sort of improvement.

The final, most general, prescription for getting better at proving things is to do lots and lots of good mathematics. The material in this book is intended to clear the decks so you can. Go to it.

4

Laboratories

The point of the following sections is to give you a chance to practice the skills from previous chapters on the sort of introductory material basic to, and often constituting the first part of, many upper-level undergraduate courses. You need to learn this material (if you don't know it already, and a more active style of reading may show you that you don't know it as well as you thought); you need to practice the skills of active reading and of proof discovery. The point of these remarks is to call to your attention this somewhat unusual double set of goals, so that you monitor your activity with respect to both of them.

4.1   Lab I: Sets by Example

The language of and standard operations on sets are part of the toolbox of every mathematician. In part this is because of their usefulness and even necessity: one wants to talk about collections of various kinds (points in the plane, functions, real numbers) all the time, and to be able to manipulate those collections in sensible ways (for example, to merge two collections into a single new collection). In part, however, the central position of sets is a holdover from the past. In the late 1800s and early 1900s the mathematical community embarked on an ambitious program to make mathematics rigorous. In particular, it was hoped that, starting with a few "primitive" or "undefined" terms and some axioms about them, one could produce all of mathematical theory from these humble beginnings. Further, it was hoped that one could prove that the resulting system was *consistent* (that is, free of contradictions) and *complete* (that is, every true theorem was provable). For most mathematicians, the undefined starting point was "set".

Early on it became clear that there were subtle obstacles in the way, and that following intuitive notions about the formation of sets created problems (see the brief discussion of Russell's paradox in Section 4.3). Only later did the whole program come screeching to a halt with the theorems of Kurt Gödel showing that the program was doomed, in the sense that consistency and completeness cannot be guaranteed for any system rich enough to do ordinary mathematics.[1] What is somewhat surprising is that, for most working mathematicians, life goes on rather as before. While it is understood intellectually that consistency isn't provable and that the theorem we're hunting may be true but unprovable, in spite of that on some psychological level we view mathematics as built up from sets in some logical fashion and tend to behave as if everything were really all right.

We won't do axiomatic set theory here; a standard reference for a somewhat more formal set theory than that introduced here (but still a presentation aimed at a general mathematician, not a student of axiomatic set theory) is Halmos' *Naive Set Theory* [12]. We do take as undefined notions *set* and *member*, and form sentences like "x is a member of X" without further inquiry. The reason for undefined terms is, of course, to avoid circularity; a diligent pursuit of any definition in the dictionary will eventually produce some circle like "A set is a collection; a collection is a family; a family is a set" and we wish to avoid this. There is a lot of duplicate language, too; we say a member of a set is an *element*, and say "is an element of" instead of "is a member of," and use the words *family* or *collection* as synonyms for sets. The notation for "is an element of" is "$\in$," so $x \in A$ means x is an element of A.

We wish a set to be determined by its members, and the way to do this is to declare two sets to be equal exactly when they have the same members. (We'll call it a definition, although more properly it is an axiom.)

Definition 4.1.1 *The sets A and B are <u>equal</u> and we write $A = B$ if every member of A is a member of B and every member of B is a member of A.*

We write $A \neq B$ if A and B are not equal.

4.1:

It follows easily that one way to describe a set completely is to list its

---

[1]The story is too complicated to go into here but makes fascinating reading. Three possible references are [8], [9], and [10]. A delightful one, if you can obtain it, is [11].

members, and the standard notation is to enclose that list in left and right set brackets "{" and "}." So, for example, $\{1, 2, 3\}$ is a familiar set.

4.2:

Since what is important is membership, the order of the listing and repetitions within the listing are unimportant.

4.3:

The production of sets by listing is feasible only for small sets. Since we are working informally, it should be no surprise that we cheat a little and write sets in which the list is only indicated and not really written, such as $\{1, 2, 4, 8, 16, \ldots, 1024\}$. Goodwill on the part of author and reader usually works just fine.

4.4:

This way of describing sets is clearly inadequate for large and/or complicated sets, and so one is allowed as well to build sets by giving a condition for membership: everything that satisfies the condition is in the set, and everything that does not is not. (It might be well to recall from Section 1.3 the informal discussion of condition.) The notation has the form "$\{x : \text{(condition on } x) \}$." So, for example, $\{x : x \text{ is an even integer } \}$ is another familiar set. Of course, there is nothing special about the letter x.

4.5:

Of course, two sets might be equal, but there are many other possibilities, as well as other sets we might wish to form from the original two. Below are some definitions and theorems that deserve thorough exploration.

Definition 4.1.2 *Given sets A and B we say A is a subset of B (or A is contained in B) and write $A \subseteq B$ if every element of A is an element of B.*

Theorem 4.1.3 *For any sets A and B, $A = B$ if and only if $A \subseteq B$ and $B \subseteq A$.*

Definition 4.1.4 *Given sets A and B we define their intersection $A \cap B$ to be the set of all elements in both A and B.*

Definition 4.1.5 *Given sets A and B we define their <u>union</u> $A \cup B$ to be the set of all elements either in A or in B.*

(Recall for this last definition that the use of "or" is inclusive (see Section 3.2).)

4.6:

Your preliminary explorations done, there are two points to be made. First, it is crucial to make clear the distinction between "is an element of" ($\in$) and "is contained in" ($\subseteq$). This is deceptively easy with basic examples of sets: nobody is likely to say $1 \subseteq \{1, 2, 3\}$, and although the incorrect $\{1\} \in \{1, 2, 3\}$ is more common it is still easy to avoid. But it is possible, and turns out to be necessary, to talk about sets whose elements (or some of whose elements) are themselves sets. One way to work through this confusion once and for all is to consider the following set:

$$\{1, \{1\}\}.$$

Determine all objects that are elements of this set and all sets that are subsets of it, and understand clearly why. For one of these tasks you will need the set with no elements at all (the <u>empty</u> or <u>null</u> set), denoted $\emptyset$.

4.7:

The second point to be made is that there is a standard way to draw pictures of sets called Venn diagrams (which you may already have known and used above). For two general sets A and B or three sets A, B, and C the diagrams are as below:

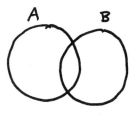 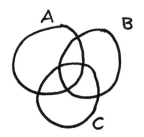

Note that in the second picture all possibilities of membership in none, one, any two, or all three of the sets have an associated area. Think a little about what diagram would be needed for four sets.

4.8:

(By the way, we have moved without saying so explicitly to considering intersections or unions of more than two sets. Think for a second about how to make these definitions for four sets, say.)

4.9:

Finally, make sure that you have explored the definitions of union, intersection, and subset using Venn diagrams. (The exercises will contain more definitions to play with, but what we have is enough for now.)

4.10:

One important aspect of sets is that they are *unordered* collections, but ordered pairs are essential for a lot of mathematics and if one really wants to get to all of mathematics starting from sets this transition must be made. Recall that the defining property of the ordered pair $(\cdot, \cdot)$ is that $(a, b) = (c, d)$ if and only if $a = c$ and $b = d$. In particular, $(3, 5) \neq (5, 3)$, and so defining the ordered pair (a, b) by $(a, b) = \{a, b\}$ is doomed to failure. Take a little while to try some ideas of your own (in part so that when you see the trick, you are properly impressed).

4.11:

If you weren't successful, don't worry — most people aren't. The following definition deserves admiration for its ingenuity.

Definition *Define the <u>ordered</u> <u>pair</u> (a, b) by $(a, b) = \{\{a\}, \{a, b\}\}$.*

(It was on the tip of my tongue too.) What needs to be verified is that this complicated looking thing actually satisfies the needed property about when equality of two ordered pairs holds. One direction is easy; the other is a good exercise in containments and proof by cases but requires care.

4.12:

This definition in hand, it is natural to form the collection of all ordered pairs with first element in some set A and second in some set B.

Definition 4.1.6 *Given sets A and B, define their <u>Cartesian</u> <u>product</u> (denoted $A \times B$) to be the set*

$$A \times B = \{(a, b) : a \in A \text{ and } b \in B\}.$$

4.13:

4.1.1 Exercises

4.14: Here's a definition to explore:

Definition *Given sets A and B, define their <u>difference</u> $A - B$ by $A - B = \{x : x \in A \text{ and } x \notin B\}$.*

Can you use this new set to express A as a union of two sets involving B? Can you express $A \cup B$ or $A \cap B$ in some interesting ways?

4.15: Sometimes all the objects under discussion are members of some fixed large set. In this case one sometimes considers a <u>universal set</u> U, and then one can define the <u>complement</u> of some subset A of U (denoted A') by $A' = U - A$. Explore, including finding some formulas involving union, intersection, and complementation. For example, what is the complement of $A \cup B$?

4.16: The following definition is sometimes useful.

Definition *Given sets A and B, define their <u>symmetric difference</u> $A \Delta B$ by $A \Delta B = (A - B) \cup (B - A)$.*

4.17: Exercise your ingenuity to produce a definition of an <u>ordered triple</u> (a, b, c), with the property, of course, that $(a, b, c) = (d, e, f)$ if and only if $a = d$, $b = e$, and $c = f$.

4.18: This exercise concerns the problem of how to count the number of elements in a union. Let's denote the number of elements in the set A by $\#A$. Convince yourself first that the formula $\#(A \cup B) = \#A + \#B$ is false in general. Since this formula sometimes works, though, it is natural to try to throw in a fudge factor to fix it. Its difficulty is that some elements are counted twice (once as a element of A and once as an element of B) by the right hand side of the proposed formula. We would like to subtract off this overcounting. The number of elements in what set is the right thing to subtract?

Now pass on to consideration of $\#(A \cup B \cup C)$. Begin with the formula $\#(A \cup B \cup C) = \#A + \#B + \#C$ (false in general), and throw in fudge factors as before. But more care and ingenuity is required! What we are playing with is some simple cases of the law of Inclusion–Exclusion, a part of the branch of mathematics called combinatorics.

4.2   Lab II: Functions by Example

If you are like many students, even students of mathematics, your notion of "function" is best described as a mess. This stems mostly from the fact that you have been given, at various times, lots of non-definitions of function that were, in the opinion of some teacher, good enough for the particular task at hand. One standard approach is to define a function as "a rule or correspondence that associates with each element of the domain one and only one element of the range," which is a non-definition because rule and correspondence are both undefined. Another approach is to lead you to believe, explicitly or implicitly, that functions are things given by formulas, so $f(x) = x^2$ is a completely typical function. You may have seen "function machines" or "input/output devices." Whatever the approach, the lack of an adequate definition was almost certainly obscured by a horde of examples from which you were supposed to figure out what a function was even if we didn't quite tell you. This was probably accompanied by a lot of hand waving and a good deal of smiling (when teachers smile too much, we're usually hiding something). Along the way was introduced some notation whose meaning is blurred, such as the difference between "f" and "$f(x)$": which is correct, "the function f" or "the function $f(x)$," and does it matter? And, almost worst of all, you have probably seen a correct definition of function comparatively recently; psychologists study with great glee the cognitive difficulties of a person with several contradictory definitions of a concept, such as which definitions are selected for use in what situations, what happens when two are in use at the same time and they conflict, and so on.

We will give in this section a correct definition of function. The major task, with you as a forewarned and active participant, is to reconcile with the proper definition those parts of what you already have that are at least pieces of the truth, and to expose and eliminate those that are really troublesome. This places on you a special burden of active exploration.

Definition 4.2.1 *Given sets A and B, a _function_ f from A to B is a subset of A × B (that is, a set of ordered pairs with first element from A and second from B) with the property that each element a ∈ A occurs as the first member of an ordered pair in f exactly once (that is, once and only once). The set A will be called the _domain_ of f and denoted domain(f), and the set B the _codomain_ of f and denoted codomain(f).*

Please observe that this definition makes a function a *set*. Pretending for a moment that this was the definition of something not called "function," explore thoroughly.

 4.19:

There's another familiar notion that deserves exploration.

Definition 4.2.2 *Given a function f from domain(f) to codomain(f), the* <u>range</u> *of f (denoted range(f)) is the subset of codomain(f) consisting of those elements that occur as the second element of some ordered pair in f.*

4.20:

Let's also note that if you have read previous discussions of relations (see Exercise 1.20 and following, among others) or functions (see Exercise 1.65 and following), it will be easy to recognize this as a special sort of relation. If you haven't read these no harm will be done; if you have, you might look back at them now.

4.21:

We now begin the process of reconciliation with the past. First, you may be used to using f to stand for a function, but not thinking of it as a set of ordered pairs. You may gain a little perspective on this by finding exactly what noun f really is in your working definition. Next, you are used to notation like "$f(x)$" or "$f(2)$"; where does this come from? The following (notational) definition makes this precise.

Definition 4.2.3 *Suppose f is a function from domain(f) to codomain(f). For each element x of domain(f), define f(x) to be the (unique) element of codomain(f) such that $(x, f(x)) \in f$.*

Translate each of the examples from your exploration above into this notation.

4.22:

Here's another formulation of function you probably encountered. Early in your mathematical career, perhaps when you were first learning to graph functions, you may have run into a function as specified by a table of values:

x	$f(x)$
a	2
b	3
c	2

(we used this device for specifying "small" functions in Section 1.6.1). A little thought will convince you that this is a presentation extremely close to the formal definition using ordered pairs.

4.23:

However, there is one catch. You probably used a table of values approach in graphing even when the function in question had a very large domain (say, all real numbers — remember how you used to graph functions by plotting points and playing connect the dots?). It is vital to realize that the tables you produced then corresponded to a few of the ordered pairs in the proper definition, and that the function was much more than the few pairs you chose to write down. The function is the set of *all* its ordered pairs, not some few you happened to select as being easy to compute. But with this understood, the table of values approach is practically identical to the ordered pairs definition. To nail this down, what would the table of values and set of ordered pairs be for the usual square function you described as $f(x) = x^2$?

4.24:

The discussion above moves us to another obstacle, which is how to reconcile the "function as formula" with "function as set of ordered pairs." To cope with this we first have to repair some damaged language. Many mathematicians, and hordes of students, write things like "the function $f(x) = x^2$." This language is wrong, because "$f(x)$" is not a function, it is the value of the function at x, or, in ordered pair language, the second element of the ordered pair in the function whose first element is x. The language is really an abbreviation for "the function f given by $f(x) = x^2$ for all x in $domain(f)$." The symbol "f" denotes the function, and "$f(x)$" denotes the value of f at x. When this language is used properly, it puts you in the following funny position: even if you have a function that you wish to call f, which is given by a formula, you cannot tell me about the function using *only* the name f, because there is no way to write down the needed formula. Try it.

4.25:

Using the language carefully places stress on the following crucial fact about functions:

To *define a function* is to *specify its values* completely.

To say that a function f is specified by a formula is then simply to say that you have a way to compute the value $f(x)$ (that is, the value of f

at x) via some formula. This can be said equally well in terms of ordered pairs; suppose the formula you have in mind gives the value of f at x to be $f(x) = x^2 + 3x + 1$, and that you have in mind that the domain is the set of real numbers. The ordered pair definition is then as follows:

$$f = \{(x, y) : x \in \mathbf{R} \text{ and } y = x^2 + 3x + 1\}.$$

Or, even better, but hard to see the first time,

$$f = \{(x, x^2 + 3x + 1) : x \in \mathbf{R}\}.$$

Do some good practice of moving from the formula definition of some functions to the ordered pair definition and *vice versa*, and work hard at writing sentences in which the distinction between f (the name of the function) and $f(x)$ (the value of the function at some point) is clear.

4.26:

Aside

At this stage a small internal voice is probably grumbling, "I've gotten along fine without worrying about all these details, and so what if I called the function $f(x)$? Nobody was upset except the professor! Does it really make any difference?" A fair answer is probably to say that you have gotten along (perhaps with more confusion and obstacles than you recognized), but mostly because most of the functions you were working with had the real numbers as both domain and codomain. In upper-level mathematics you will have, sooner or later, to talk about functions whose domain and range are other things, perhaps even themselves sets of functions. If you don't have the distinction between function and value clearly in mind, and expressed with appropriate notation, this is impossible. And there are other situations in which this confusion, even if not deadly, will be frustrating and inhibiting even if you can't quite identify what the problem is.
End Aside

With the past and present in agreement (we hope), here are some definitions to explore. They are given in terms of the ordered pair notation for functions, but after you have explored them this way, you are strongly encouraged to put them into the $f(x)$ notation and explore them all over again.

Definition 4.2.4 *A function f is injective (alternatively, one-to-one) if each element of codomain(f) is the second element of at most one ordered pair in f.*

We sometimes say an injective function is an injection.

Definition 4.2.5 *A function f with codomain B is* <u>surjective</u> <u>on</u> *B (alternatively,* <u>onto</u> *B) if each element of B occurs as the second element of an ordered pair in f at least once. If the codomain B is understood, we sometimes simply say that f is* <u>surjective</u> *or* <u>onto</u> *or that it is a* <u>surjection</u>.

Definition 4.2.6 *A function f with domain A and codomain B is said to be a* <u>bijective function</u> *between A and B (alternatively, a* <u>one-to-one correspondence</u> *between A and B or a* <u>bijection</u> *between A and B) if it is both injective and surjective on B.*

4.27:

There are several ways to draw pictures to help understand functions. One is the sort of diagram used in Section 1.6.1 and reproduced below:

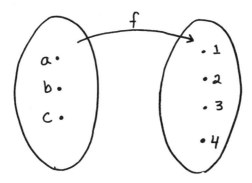

We may as well deal with the graph of a real-valued function as well. Observe that the graph of a real-valued function f is simply a plot of all the ordered pairs in f. Explore a little with these two sorts of pictures.

4.28:

These tools in hand, we may return to definitions.

The definition of composition of functions generally is troublesome when first encountered, but the ordered pair definition of function actually makes things easier.

Definition 4.2.7 *Suppose f and g are functions satisfying the equation* $range(f) \subseteq domain(g)$. *Then the function* $g \circ f$ *(the* <u>composition</u> *of g and f) is the function*

$$g \circ f = \{(a, c) : \text{ there exists } b \text{ so that } (a, b) \in f \text{ and } (b, c) \in g\}.$$

After you are comfortable with this on its own merits, one part of your exploration ought to be to reconcile this approach to composition with the more familiar notation. In particular, what are "c" and "b" in terms of a in the usual notation? Also, make sure to draw some pictures.

4.29:

There's another definition that the ordered pair formulation of function actually makes easier for most people, which is the definition of inverse function. Recall that, intuitively, the inverse of a function f from A to B is some function that goes from B to A and "goes backward" or "undoes the action of f." Since a function is a set of ordered pairs in $A \times B$, surely if we simply assemble the collection of all the "reversed" ordered pairs in f we will at least get something of the right general sort. Explore this for a while, with two questions in mind: does the object you get by doing this seem to be the right candidate for an inverse, and what conditions do you need on f to make sure that the object you get is actually a function?

4.30:

You now ought to be able to write a proper definition of inverse function, with the appropriate conditions on f to ensure it actually has an inverse function. Recall that the inverse of f, if it exists, is denoted f^{-1}.

4.31:

Finally, does the object you defined have the desirable (actually, essential!) properties that $f^{-1} \circ f$ is the identity function on $domain(f)$ and $f \circ f^{-1}$ is the identity function on $range(f)$? (The identity function on a set S is the function $i_S = \{(x, x) : x \in S\}$, which may need its own exploration.)

4.32:

4.2.1   Exercises

4.33: Suppose you are given sets A and B and a function f from A to B. There is also a set C and some function h from A to C on which you get to put some condition or conditions; the question is, what condition is required on h so that you may (always, no matter what f is) construct a function g from C to B so that $f = g \circ h$? Explore with examples (pictures are useful) to come to a conjecture.

4.34: Suppose you are given sets A and B and a function f from A to B. There is also a set C and a function g from C to B on which you get to put some condition. The goal is to make a guess as to what property g must have so that (for any f) you may always find a function h from A to C so that $f = g \circ h$.

4.35: Suppose you are given nonempty sets A and B. What does a function with domain $A \times B$ and range A look like (be careful!)? What about range $A \times B$ and domain some set C? What about a function from $A \times B$ to $C \times D$? These in hand, construct a surjective function from $A \times B$ to A. Construct an injective function from A to $A \times B$. Construct a bijection between $A \times B$ and $B \times A$. In these cases, "formulas" really are possible and desirable.

4.36: We consider in this problem the collection of functions from $\mathbf{R}$ to $\mathbf{R}$, where $\mathbf{R}$ denotes the real numbers. Define addition on this collection by, given f and g, defining $f + g$ by $(f + g)(x) = f(x) + g(x)$ for all $x \in \mathbf{R}$. Translate this definition into ordered pair notation and explore a little. Can you find an identity element and inverses for this addition?

4.37: In the past you may have talked about the <u>restriction</u> of a function; using the language of that time, you "kept the same rule/formula but used a different (smaller) domain." Suppose we have a function f from A to B, and we wish to consider its restriction to some different domain $A_1 \subseteq A$. It turns out that with the definition of function as "set of ordered pairs," this restricted function is expressible as a simple intersection of two sets. Express it!

4.3   Lab III: Sets and Proof

The purpose of this section is to use formal language, especially quantifiers, to make precise the definitions of various set operations. The definitions given in Section 4.1 talked around quantifiers, and that's good enough for some purposes but not for proofs. For example, the <u>set</u> <u>builder</u> <u>notation</u> described there, in which a set is specified by a condition, says really that one may build sets by forming

$$\{x : P(x)\},$$

where $P(x)$ is a statement form in x.[2] Observe also that there is an implicit universal quantifier, which actually shows up if the notation is read aloud:

---

[2]This is still imprecise: P must be a statement form in which x is the only "free variable," so something like $\{x : x = y^2\}$ is disallowed because y is another variable. Something like $\{x : \exists y(x = y^2)\}$ is allowed, because y is a "bound" variable (bound by the quantifier $\exists$). We won't go into bound and free variables

the above is the set of *all* x satisfying $P(x)$, not just some of them.

This device of forming a set by selecting all things satisfying a condition (or making a statement form true, or having a property — same thing) is extremely natural. We want to be able to form the set of all odd numbers, or all continuous real-valued functions, and so on, and it seems reasonable to do so. But it turns out that allowing this much power by being able to use *any* P creates difficulties. Russell's Paradox is an example of what can go wrong. Consider the statement form P given by $x \notin x$. This is a statement form in the one variable x, and although it seems unusual that a set could possibly be an *element* of itself the statement form is perfectly fine. Well, we may then form the set S from this condition by $S = \{x : x \notin x\}$. A suspicious mind then might ask, is $S \in S$? Or is $S \notin S$? Surely one or the other must be true. Examine both possibilities until you see that each leads, directly from the definitions, to a contradiction (hence the paradox).

4.38:

This paradox (introduced by Bertrand Russell in 1901) and some others exposed some of the obstacles mentioned in Section 4.1 to a straightforward and *consistent* development of set theory. It turns out that crafting a set of axioms for building sets that allows you to get the sets you want but excludes objects like the one above is delicate. There are various approaches, and the history makes good reading. One approach is to allow this sort of formation of a set by a condition as long as what you really form is the *subset* of a *known* set satisfying the condition. You then give yourself some fairly conservative rules for building up a family of sets you know are safe, and then off you go (the reader is again referred to Halmos' *Naive Set Theory* [12]). Of course, Gödel's theorems later showed that you couldn't guarantee consistency anyway, but the workers of the time were at least able to build (various) systems that excluded known paradoxes. We won't worry further about these matters, since it is likely that none of the sets you will actually try to form in practice really encounters these difficulties.

Given a set A the expressions $x \in A$ and $x \notin A$ are clearly statement forms. Therefore, the following formal versions of the definitions in Section 4.1 are immediate.

Definition 4.3.1 *We say two sets A and B are* <u>*equal*</u>*, and write $A = B$, if $\forall x(x \in A \Leftrightarrow x \in B)$.*

Definition 4.3.2 *Given sets A and B we say A* <u>*is a subset*</u> *of B (or A is* <u>*contained in*</u> *B) and write $A \subseteq B$ if $\forall x(x \in A \Rightarrow x \in B)$.*

___

here, but a rule of thumb test for a legal P is to ask whether the substitution of any allowed value for x results in a statement, and not just a statement form.

Definition 4.3.3 *Given sets A and B we define their <u>intersection</u> $A \cap B$ by $A \cap B = \{x : x \in A \text{ and } x \in B\}$.*

Definition 4.3.4 *Given sets A and B we define their <u>union</u> $A \cup B$ by $A \cup B = \{x : x \in A \text{ or } x \in B\}$.*

Definition 4.3.5 *Given sets A and B, define their <u>Cartesian product</u> (denoted $A \times B$) to be the set*

$$A \times B = \{(a,b) : a \in A \text{ and } b \in B\}.$$

(More definitions are to be found in the exercises in Section 4.1.1; their quantified forms are already displayed there.)

We'll shortly consider expressions containing mixed unions and intersections of sets, which raises the usual "order of operations" questions: should $A \cap B \cup C$ be interpreted as $A \cap (B \cup C)$ or $(A \cap B) \cup C$? We will simply avoid this difficulty by using parentheses so no ambiguity results.

The exercises following give some of the many results of what is sometimes called the algebra of sets; other results will suggest themselves from these. All deserve exploration, especially with Venn diagrams, as well as proof.

4.3.1 Exercises

4.39: Prove that $A = B$ if and only if $A \subseteq B$ and $B \subseteq A$. It is worth remarking that this is a result often (very, very often) used to prove that two sets are equal.

4.40: (Distributive Laws) Prove that $A \cap (B \cup C) = (A \cap C) \cup (A \cap C)$ and $A \cup (B \cap C) = (A \cup B) \cap (A \cup C)$.

4.41: (De Morgan's Laws) Prove that $A - (B \cup C) = (A - B) \cap (A - C)$ and $A - (B \cap C) = (A - B) \cup (A - C)$.

4.42: Prove $A \Delta B = \emptyset$ if and only if $A = B$.

4.43: Prove or disprove each of the following:

i) $A \cup B = A \cup C$ implies $B = C$,

ii) $(A \cap B = A \cap C$ and $A \cup B = A \cup C)$ implies $B = C$,

iii) $A \cup B \subseteq A \cap B$ implies $A = B$,

iv) $A \subseteq B$ if and only if $A \cup B = B$.

4.44: Prove that $(A \cup B) \times C = (A \times C) \cup (B \times C)$. Determine and prove at least one more relationship involving set operations and Cartesian products.

4.45: Here's a definition of a set that careful axioms of set formation do allow you to form (and know is a set).

Definition 4.3.6 *Suppose A is a set. Define the <u>power set</u> of A, which we denote by $\mathcal{P}(A)$, by*

$$\mathcal{P}(A) = \{X : X \subseteq A\}.$$

In dealing with power sets, it is worth holding tightly to the fact that $X \in \mathcal{P}(A)$ if and only if $X \subseteq A$.
   Prove the following:

i) $A \subseteq B$ if and only if $\mathcal{P}(A) \subseteq \mathcal{P}(B)$,

ii) $\mathcal{P}(A) \cup \mathcal{P}(B) \subseteq \mathcal{P}(A \cup B)$.

   Determine the relationship between $\mathcal{P}(A) \cap \mathcal{P}(B)$ and $\mathcal{P}(A \cap B)$ and prove it.
   Prove that $\mathcal{P}(\emptyset) = \{\emptyset\}$.

4.46: Prove that if the number of elements of a set S is (a positive integer) n, then the number of elements of $\mathcal{P}(S)$ is 2^n.

4.4   Lab IV: Functions and Proof

Recall that we defined a function f from A to B to be a subset of $A \times B$ with the special property that each element a of A occurs exactly once (once and only once) as the first member of an ordered pair in f. This is perfectly correct and is adequate for constructing examples. It turns out to be hard to use to prove things, though, since "exactly once" is in fact a way of avoiding the use of quantifiers; if we make these explicit, proofs turn out to be easier to do mechanically.
   In uncovering the hidden quantifiers it is useful to separate "exactly once" into "once and only once" and deal with each piece separately. Let's start with "once"; how can you use quantifiers to say that each element of A occurs once as the first member of an ordered pair in f?

4.47:

If that seems tricky, it's because it is. In English when we say "once" we really usually mean "once and only once" (when I say once, I don't mean twice). What turns out to work here is to turn "exactly once" into "(at least once) <u>and</u> (at most once)." It is confusing but true that mathematicians use "once" for the first of these and "only once" for the second. So the first task is really to figure out how to say that every a in A occurs at least once as the first member of an ordered pair in f.

4.48:

It is even harder to figure out how to say "only once," meaning "at most once," using quantifiers unless you have seen the trick. The way somebody clever figured out how to do this was really to say the following: if it appears to happen twice, then the two apparently different occurrences were actually the same. Take a shot at expressing, using quantifiers, that every a in A occurs as the first element in an ordered pair in f at most once.

4.49:

It isn't easy, but at least it can be done. We should note along the way that this is exactly the strategy for expressing the idea that something "exists and is unique." To say that something exists is to say that there is at least one. While it isn't quite normal English to say that "unique" means "at most one" (when was the last time you said something was unique when you knew there weren't any?), it is true that the *combination* of "at most one" and "at least one" gives you the result "exists and is unique." This may even have been the definition of function as set of ordered pairs you saw: a set of ordered pairs in which every element of the domain occurs as the first element of a unique ordered pair in f.

Aside

This trick appears frequently and is worth remembering. For example, you learned long ago that limits were unique if they existed. Just for practice, figure out how to express that using quantifiers.

4.50:

Also, you may see the proof in Exercise 2.18 for another example of this trick. Finally, you might compare this with the approach used in the Hint for Exercise 3.27 to say that there exist distinct (things).
End Aside

We may finally give a careful definition of function with all quantifiers fully displayed.

Definition 4.4.1 *Given sets A and B, a <u>function</u> f from A to B is a subset of $A \times B$ with two properties:*

i) $\forall a(a \in A \Rightarrow \exists b((a, b) \in f))$, *and*

ii) $\forall a((a \in A \text{ and } (a, b) \in f \text{ and } (a, c) \in f) \Rightarrow b = c)$.

The set A will be called the <u>domain</u> of f and denoted domain(f), and the set B the <u>codomain</u> of f and denoted codomain(f). The set $\{b : \exists a((a, b) \in f)\}$ will be called the <u>range</u> of f and denoted range(f).

Remark that if we use notation like "$\forall a \in A(\ldots)$" these can be expressed more neatly.

4.51:

The next task is to turn all this into the standard function notation involving $f(x)$.

4.52:

The two other standard definitions coming along with that of function are <u>injective</u> and <u>surjective</u>, which were given in correct but unquantified form in Section 4.2. Give their forms with quantifiers fully displayed (in both ordered pair and "function" notation).

4.53:

It's worth stressing that to prove a function is surjective you are inevitably in an existence proof and will have to produce something; to prove a function is injective you will have to assume that two things are present and show they actually coincide.

The definitions of composition of functions and inverse function are so straightforward we'll save you the trouble and write them down.

Definition 4.4.2 *Suppose f and g are functions satisfying the equation range(f) ⊆ domain(g). Then the function g ∘ f (the <u>composition</u> of g and f) is the function*

$$g \circ f = \{(a, c) : \exists b((a, b) \in f \text{ and } (b, c) \in g)\}.$$

Definition 4.4.3 *Suppose f is a bijection from A to B. Then the set $\{(b, a) : (a, b) \in f\}$ is a function from B to A (in fact, a bijection), which we call the <u>inverse</u> of f and denote f^{-1}.*

(WARNING: did you check that the given set actually did possess the properties needed to be a function? If not, you stand convicted of lazy reading.)

4.54:

The first of the exercises below is to prove that this f^{-1} has the required properties to be the "inverse."

4.4.1 Exercises

4.55: Suppose f is a bijection from A to B, and f^{-1} is defined as above. Then $\forall a \in A((f^{-1} \circ f)(a) = a)$ and $\forall b \in B((f \circ f^{-1})(b) = b)$.

4.56: We have defined a function as a set; we have defined when two sets are equal (see Section 4.3); we therefore have defined when two functions are equal. Prove that our definition coincides with the following: two functions f and g are equal if $domain(f) = domain(g)$ and $\forall x \in domain(f)(f(x) = g(x))$. Please note that "same formula" appears nowhere in here, nor is it going to.

4.57: A generalization of the idea of function from A to B is that of a <u>relation</u> from A to B. We give the definition here, but note that you've seen this definition before in problems from Section 1.2.1 and several sections following.

Definition 4.4.4 *A <u>relation</u> from A to B is a subset of $A \times B$.*

(If A and B are the same, we sometimes say we have a relation "on A.") There are four conditions one might put on a general relation R:

1. Each element of A appears as the first element of at least one ordered pair in R,

2. Each element of A appears as the first element of at most one ordered pair in R,

3. Each element of B appears as the second element of at least one ordered pair in R,

4. Each element of B appears as the second element of at most one ordered pair in R.

Which conditions on a relation guarantee it is a function? An injective function? A surjective function? How would you define a <u>surjective</u> <u>relation</u>? An <u>injective</u> <u>relation</u>? Here's one more definition.

Definition 4.4.5 *Given a relation R from A to B, the relation given by $\{(b, a) : (a, b) \in R\}$ is called the <u>inverse</u> <u>relation</u> for R and denoted R^{-1}.*

Which conditions on a general relation R guarantee that R^{-1} is an injective relation? That R^{-1} is a function? That both R and R^{-1} are functions? Give proofs of your answers.

Finally, define the composition of relations and find, with proof, $R \circ R^{-1}$.

4.58: Suppose f and g are functions such that the composition $g \circ f$ is defined. Prove that if f and g are injective then $g \circ f$ is injective, and that if f and g are surjective then $g \circ f$ is surjective. Thus injectivity of the individual functions is sufficient for injectivity of the composition; is it necessary? What about for surjectivity?

4.59: Suppose that we are considering subsets of some fixed universal set U.

Definition 4.4.6 *Let A be a subset of U. The function χ_A with domain U and codomain $\{0, 1\}$ is defined by*

(4.1)
$$\chi_A(x) = \begin{cases} 0, & x \notin A, \\ 1, & x \in A. \end{cases}$$

The function χ_A is called the __characteristic function__ of A.

Note that this definition really defines a whole family of functions with common domain U, one function for each subset of U. Prove that $A = B$ if and only if $\chi_A = \chi_B$. Prove that $\chi_{A \cap B} = \chi_A \cdot \chi_B$.

4.60: Suppose that A and B are sets. Find an explicit bijection between $A \times B$ and $B \times A$, and prove that it is one.

4.5   Lab V: Function of Sets

Functions are defined by their values, and it is often productive to think about a function f as mapping each element in the domain A to a unique element of some codomain B. One might describe this as saying that functions act "pointwise." Sometimes, though, one wants to consider some subset S in the domain and its image under f (that is, the collection of elements of B which are $f(a)$ for some $a \in S$). Less obviously, the pre-image under f of some set C contained in B is of interest, consisting of all those $a \in A$ such that $f(a) \in C$. In fact, the generalization of "continuous" out of the setting of real-valued functions is formulated exactly in terms of the behavior of pre-images.

The above descriptions can be turned into actual definitions with a little work, and we record the formal definitions below.

Definition 4.5.1 *Let f be a function with domain A and codomain B. For each subset S of A, define $f(S)$ by*

$$f(S) = \{b \in B : \exists a(a \in S \text{ and } b = f(a))\}.$$

We call the set $f(S)$ the __image__ of S under f.

Definition 4.5.2 *Let f be a function with domain A and codomain B. For each subset S of B, define $f^{-1}(S)$ by*

$$f^{-1}(S) = \{a \in A : \exists b(b \in S \text{ and } b = f(a))\}.$$

We call the set $f^{-1}(S)$ the pre-image of S under f.

It is worth checking that shorter forms (hiding quantifiers) $f(S) = \{f(a) : a \in S\}$ and $f^{-1}(S) = \{a : f(a) \in S\}$ are in fact correct, although not necessarily easier to use in proofs. Also, there are a few trivial propositions about $f(A)$, $f(\emptyset)$, $f^{-1}(range(f))$, and so on, which we leave to the reader. More important is to cope with a possible notational problem. Note that the definition of pre-image defines the whole symbol "$f^{-1}(S)$," and it is dangerous to think of this as "the function f^{-1} of S" since f^{-1} need not even exist. However, if f does happen to be invertible, "$f^{-1}(S)$" has two legitimate interpretations: one is as defined in Definition 4.5.2, which does not use the existence of f^{-1}; the other one views f^{-1} as a function in its own right and uses Definition 4.5.1 as applied to f^{-1} and a subset S of its domain. You should check that the two subsets of A obtained under these interpretations coincide, thus resolving the potential ambiguity.

There are plenty of results summarizing how images and pre-images behave under the standard set operations; we summarize a few below, leaving the proofs to the reader.

Proposition 4.5.3 *Suppose f is a function from A to B. Suppose that A_1 and A_2 are subsets of A, and B_1 and B_2 are subsets of B. Then*

i) $f(A_1 \cup A_2) = f(A_1) \cup f(A_2)$,

ii) $f(A_1 \cap A_2) \subseteq f(A_1) \cap f(A_2)$,

iii) $f^{-1}(B_1 \cup B_2) = f^{-1}(B_1) \cup f^{-1}(B_2)$,

iv) $f^{-1}(B_1 \cap B_2) = f^{-1}(B_1) \cap f^{-1}(B_2)$,

v) $A_1 \subseteq f^{-1}(f(A_1))$,

vi) $f(f^{-1}(B_1)) \subseteq B_1$,

vii) $f^{-1}(B_1) - f^{-1}(B_2) = f^{-1}(B_1 - B_2)$,

viii) $f(A_1) - f(A_2) \subseteq f(A_1 - A_2)$.

There is another way to view these definitions, which is to regard them as definitions of some new functions. Some books call these the *induced set functions* and we can give their definitions in terms of those already made.

Definition 4.5.4 *Suppose f is a function from A to B. We define functions F from $\mathcal{P}(A)$ to $\mathcal{P}(B)$ and F_n from $\mathcal{P}(B)$ to $\mathcal{P}(A)$ by*

$$F(S) = f(S), \quad S \in \mathcal{P}(A),$$

and

$$F_n(S) = f^{-1}(S), \quad S \in \mathcal{P}(B).$$

Before we go further, make sure that you understand clearly the domains, ranges, and definitions of F and F_n.

The reason for finding some solid ground is because some books take things a step further and use the same name f for both the original function and the induced function we have called F, and use f^{-1} for what we have called F_n. This actually doesn't create problems if one thinks carefully but requires context to make clear which use of these symbols is meant. For example, the statement $\forall S(f(S) = f(S))$ is *not* trivial if the "$f(S)$" on the left-hand side of the equation is interpreted as that meant by Definition 4.5.1 while the right-hand "$f(S)$" is interpreted with f viewed as the induced set function from $\mathcal{P}(A)$ to $\mathcal{P}(B)$. The following questions might help nail this down: under what interpretations is f a function? What about f^{-1}? Which of the following are always defined, and which only sometimes: $f(a)$, $f(\{a\})$, $f^{-1}(b)$, $f^{-1}(\{b\})$?

We'll continue to use the original notation for these induced set functions. The question to be asked is, how are properties of f and F related? Injectivity, surjectivity, and bijectivity are all fair game. Also, one can fairly ask how F and F_n behave with respect to the ordinary set operations (containment, union, and so on). None of the conjectures or proofs is particularly hard, but they give good practice in formal proofs and the use of definitions.

4.5.1   Exercises

4.61: The first and most important exercise for this section is to critique your reading of the section you just finished. There were no icons to interrupt your reading when it was time to do something, so that awareness was up to you. Did you explore with examples? Did you stop and prove, or try to prove, things "left to the reader"? The point is that we are running out of chances to get you to change your reading style on a permanent basis, so you need to take over the burden. The following is a *minimal* list of places you ought to have stopped and done some active work.

CALL TO ACTION	ACTION NEEDED
The first paragraph of informal definitions	Example exploration
"can be turned into formal definitions"	Effort at the formal definitions
The formal definitions	Thorough example exploration
Shorter forms of the definitions	Example exploration

CALL TO ACTION	ACTION NEEDED
"a few trivial propositions"	Explorations and proofs and conjectures (for example, $f^1(\emptyset)$ is not in the list, but you should have noticed that and considered it)
"you should check"	Actually prove that $f^{-1}(S)$ is the same under the two interpretations
"leaving the proof to the reader"	Prove all of the items in the proposition and, further, explore by example those that are containments instead of equalities to see why
Definitions of induced set functions	Exploration, especially thorough in view of "Before we go further ..."
The questions in the paragraph on alternate notation for induced set functions	Answers(!)
Questions on the properties of f and F	Exploration, list of conjectures, proofs

(Notice that if you actually prove all the things in the proposition and all other results suggested, no exercises in these concepts are really needed. The next exercise is solely to give you a chance to see whether your collection of conjectures about induced set functions was big enough, and whether you were thinking when you made them.)

4.62: Turn each of the results in the proposition into the notation of induced set functions and prove a few in that language. Generate, and then prove, a correct conjecture about how induced set functions behave on the symmetric difference of sets $S_1 \Delta S_2$. Think a little before you prove it; there are two approaches.

4.6   Lab VI: Families of Sets

[A WORD OF WARNING: you are on your own for active reading. Show your stuff.]

It is frequently necessary in mathematics to consider some set whose elements are themselves sets. To try to indicate which objects are viewed as elements, it is common to say a *collection* of sets or a *family* of sets. There's also often an effort to find a notational distinction between the sets that are elements and the big set, like script letters "$\mathcal{S}$" for the big set and "A," "B," etc., for the elements. Examples abound: since any function is a set, a collection of functions is a family of sets; since a line is a set of points, a family of lines is a family of sets; in topology, the family of what are called open sets is crucial. We discuss in this section various elaborations of this basic idea and the operations needed to work with a family of sets.

In some sense you have already worked with a family of sets, because when you considered $A \cup B$ you might have thought you were working with a set $\{A, B\}$ (a family of just two sets), although there's no particular reason to think of it that way. It should be clear that the complications (if any) arise from considering a family that is bigger than that — perhaps even infinite. But one might still want to do the same sorts of things one did with a family of two sets, like take the union or intersection of all the sets in the family. These require some definitions.

Definition 4.6.1 *Let $\mathcal{S}$ be a family of sets. The <u>union</u> of the sets in $\mathcal{S}$, denoted $\cup_{S \in \mathcal{S}} S$ or just $\cup_{\mathcal{S}} S$, is the set*

$$\cup_{S \in \mathcal{S}} S = \{x : \exists S(S \in \mathcal{S} \text{ and } x \in S)\}.$$

Sometimes this is called the union of $\mathcal{S}$.

It is straightforward to check that if $\mathcal{S}$ is a set with two (sets as) elements, this gives the ordinary pairwise union.

You would have been well within your rights to expect the definition of intersection to accompany that of union, and what you might have expected is:

The <u>intersection</u> of the sets in $\mathcal{S}$ is denoted by $\cap_{S \in \mathcal{S}} S$ or just $\cap_{\mathcal{S}} S$ and is defined by

$$\cap_{S \in \mathcal{S}} S = \{x : \forall S(S \in \mathcal{S} \Rightarrow x \in S)\}.$$

There is, however, an annoying possibility to be considered: suppose $\mathcal{S}$ is the empty set? You should check that the union of the empty family of sets is empty: $\cup_{\emptyset} S = \emptyset$. That's fine, but manipulation of quantifiers with the above attempt at a definition for intersection appears to show that $\cap_{\emptyset} S = \{\text{everything}\}$, and this is a problem. For one thing, this would imply that $\cap_{\emptyset} S \not\subseteq \cup_{\emptyset} S$, and this is not the way that union and intersection ought to work. Much more importantly, as it turns out, as discussed in

Section 4.3, there are sets one ought not to try to build if one hopes for a consistent system, and this is one of them. It is therefore customary to leave the intersection of the empty family of sets undefined (this is rather like leaving 0^0 undefined). Thus the definition is made with this situation ruled out.

Definition 4.6.2 *Let S be a nonempty family of sets. The* <u>*intersection*</u> *of the sets in S is denoted by $\cap_{S \in S} S$ or just $\cap_S S$ and is defined by*

$$\cap_{S \in S} S = \{x : \forall S (S \in S \Rightarrow x \in S)\}.$$

We then have the following unsurprising results whose proofs are left to the reader.

Proposition 4.6.3 *Let S be any nonempty family of sets. Then*

$$\cap_S S \subseteq \cup_S S.$$

Proposition 4.6.4 *Let S be any nonempty family of sets, and let T be a nonempty subset of S. Then*

$$\cap_S S \subseteq \cap_T T,$$

and

$$\cup_T T \subseteq \cup_S S.$$

Corollary 4.6.5 *Let S be any nonempty family of sets. Then for any element S_0 of S,*

$$\cap_S S \subseteq S_0 \subseteq \cup_S S.$$

It turns out that an infinite family of sets can display some counter-intuitive behavior, and in the exercises we ask you to come up with some examples of this. In the absence of any assumptions about the elements of the family, results other than those above are hard to come by, but two important families of sets (where the members are assumed to have certain structure) are the *measurable* sets (in the theory of integration) and the *open* sets (in topology).

We turn next to another device for dealing with families of sets, which is the notion of *indexed* family of sets. We've seen this before when we considered some sets A_1 and A_2. The intuitive idea is that although both of the sets have the name "A," the subscript acts as an index to point to different sets. A similar idea is used with sequences of real numbers, where we denote a sequence s_1, s_2, s_3, ... and know that the term s_2 is not the same as s_3. In fact, if you have seen a formal definition of sequence, the following definition for an indexed family of sets will look a little familiar.

Definition 4.6.6 *An* <u>*indexed family of sets*</u> *is a nonempty family of sets S, a set Λ, and a surjective function f from Λ to S. For each α in Λ, we* *denote the set* $f(\alpha)$ *by* S_α*; note that each element of* $\mathcal{S}$ *is* S_α *for at least one* α*. Each* α *in* Λ *is called an* _index_ *and* f *is called an* _indexing function_. *We usually write* $\mathcal{S}$ *as* $\{S_\alpha : \alpha \in \Lambda\}$.

We will often just write "let $\{S_\alpha : \alpha \in \Lambda\}$ be an indexed family of sets" for ease of language. Observe that, to avoid difficulty with intersection, we assume right from the beginning that the family of sets is nonempty.

Definition 4.6.7 *Let* $\{S_\alpha : \alpha \in \Lambda\}$ *be an indexed family of sets. Define the* _union_ *of* $\{S_\alpha : \alpha \in \Lambda\}$*, denoted* $\cup_{\alpha \in \Lambda} S_\alpha$ *or just* $\cup_\Lambda S_\alpha$ *by*

$$\cup_{\alpha \in \Lambda} S_\alpha = \{x : \exists \alpha (\alpha \in \Lambda \text{ and } x \in S_\alpha)\}.$$

Define the _intersection_ *of* $\{S_\alpha : \alpha \in \Lambda\}$*, denoted* $\cap_{\alpha \in \Lambda} S_\alpha$ *or just* $\cap_\Lambda S_\alpha$ *by*

$$\cap_{\alpha \in \Lambda} S_\alpha = \{x : \forall \alpha (\alpha \in \Lambda \Rightarrow x \in S_\alpha)\}.$$

Observe that we have now defined the union of "all the sets in a family" in two potentially conflicting ways if the family happens to be indexed, once in Definition 4.6.1 (don't use the index) and once in Definition 4.6.7 (use the index). Check that there is no trouble, since both definitions yield the same set, and that the same situation occurs for the definitions of intersection.

We leave to the reader the verification that the analogues of the two propositions and the corollary above about families of sets hold when appropriately rewritten for indexed families of sets. The following results are the distributive laws (see Exercise 4.40 in Section 4.3.1 for the version of these for a family with just two sets).

Proposition 4.6.8 *Let* $\{S_\alpha : \alpha \in \Lambda\}$ *be an indexed family of sets and let* B *be a set. Then*

$$B \cap (\cup_{\alpha \in \Lambda} S_\alpha) = \cup_{\alpha \in \Lambda} (B \cap S_\alpha),$$

and

$$B \cup (\cap_{\alpha \in \Lambda} S_\alpha) = \cap_{\alpha \in \Lambda} (B \cup S_\alpha).$$

We leave the proofs to the reader but point out that there is more going on than at first appears. The left-hand side of the first equality is surely defined, but what about the right-hand side? Well, we are using implicitly that there is a second family of sets out there with the same index set Λ, namely the indexed family $\{T_\alpha : \alpha \in \Lambda\}$, where $T_\alpha = B \cap S_\alpha$ for all α. That's not so hard once pointed out, but you should be alert for other situations in which an indexed family of sets is defined in passing like this.

There's a definition that turns out to be more useful than it looks. First, it is natural to say that the (nonempty) family $\{S_\alpha : \alpha \in \Lambda\}$ is _disjoint_ if $\cap_\Lambda S_\alpha = \emptyset$. The following definition describes another possible property; you should verify that it is in fact not the same as that of being disjoint.

Definition 4.6.9 *Let* $\{S_\alpha : \alpha \in \Lambda\}$ *be an indexed family of sets. We say* $\{S_\alpha\}$ *is* <u>*pairwise disjoint*</u> *if, for each pair* α *and* β *in* Λ *with* $\alpha \neq \beta$, $S_\alpha \cap S_\beta = \emptyset$.

It should be clear why the requirement $\alpha \neq \beta$ is needed, since the only set S_α such that $S_\alpha \cap S_\alpha = \emptyset$ is $S_\alpha = \emptyset$. But this is also a good place to stress that the general definition of indexed family of sets allows *repetition* of sets in the same way that a sequence may have repeated terms. Indeed, an extreme example of this is that in which each of the S_α is some common set S. In this sense a general family of sets and an indexed family of sets are not quite the same thing; the notion of index really does yield something new. We warn the reader also that if $A_1 = \{1\}$, $A_2 = \{2\}$, $A_3 = \{3\}$, $A_4 = \{1\}$, then $\{A_1, A_2, A_3, A_4\}$ is the same family of sets as $\{A_1, A_2, A_3\}$ (because in a set, repetitions are ignored), but not the same indexed family of sets; on some occasions, it makes things easier to be a little sloppy about this, and pretend that they are the same indexed family of sets (for example, if all you care about is their union, sloppiness is not punished).

A particularly natural set to use as an index set is the set of positive integers $\mathbf{N}$. This is useful by analogy with sequences, and one strength of the analogy not perhaps obvious is that there is an *order* on the positive integers. The order is of course "$<$," and while there are various subclasses of orders with various sets of properties, "$<$" on $\mathbf{N}$ is as good as it gets. It was crucial, and correct, to be able to talk about the first term in the sequence, the second term in the sequences, and so on, and that's also useful when dealing with a family of sets. By the way, the union $\cup_n S_n$ is often written $\cup_{n=1}^{\infty} S_n$, and the notation $\cup_{k=1}^{n} S_k$ also stands for something useful.

For example, suppose we have some family of sets $\{S_n : n \in \mathbf{N}\}$ indexed by the positive integers and we rather wish that we could replace $\{S_n\}$ by another family of sets $\{T_n\}$ so that $\cup_n S_n = \cup_n T_n$ but the $\{T_n\}$ are pairwise disjoint. There's a standard way to do this: set $T_1 = S_1$ and, for each $n \geq 2$, set $T_n = S_n - (\cup_{i=1}^{n-1} S_i)$. We leave you to check that the T_n are pairwise disjoint, and that not only is $\cup_n T_n = \cup_n S_n$ but in fact the stronger property $\cup_{i=1}^{n} S_i = \cup_{i=1}^{n} T_i$ holds for all n. A problem to think about is how, given a family $\{S_n : n \in \mathbf{N}\}$, one could produce a family $\{R_n : n \in \mathbf{N}\}$ such that $\cup_{i=1}^{n} S_i = \cup_{i=1}^{n} R_i$ for all n and the sets R_n are <u>increasing</u>, in the sense that $n < m$ implies $R_n \subseteq R_m$.

4.6.1 Exercises

4.63: Produce a family of sets whose intersection is $\{0\}$ but none of which is $\{0\}$ itself. Produce an indexed family of sets, indexed by the rational numbers (fractions) whose intersection is $\{0\}$ but none of which is $\{0\}$ itself. Can you produce a finite family of sets whose intersection is $\{0\}$ but none of which is $\{0\}$ itself? Can you produce a finite family of sets whose intersection is $\{0\}$ but such that the intersection of any two of them is not $\{0\}$?

4.64: The power of having an arbitrary index set, instead of limiting the index set to the integers, say, probably isn't apparent. Consider the following task: suppose we wish to write the set of all polynomials with integer coefficients of degree three or less as the union of a family of sets, but we wish to do it in a very special way. We wish to be sure that each of the sets in the family, say S_α, is such that there is an easy, transparent bijection from S_α to the integers. Here's the way to succeed: for each polynomial p of degree two or less, let S_p be the set $\{nx^3 + p(x) : n \text{ is an integer}\}$. Explore this thoroughly with examples. Now collect all the sets $\{S_p : p \text{ has degree two or less}\}$ into an indexed family of sets (what is the index set?). Does this family have the required property? You will see this idea when you prove that the collection of all polynomials with integer coefficients is *countable* (whatever that is).

4.65: Suppose you are given two indexed families of sets $\{S_\alpha : \alpha \in \Lambda\}$ and $\{T_\beta : \beta \in \Gamma\}$. You wish to consider the collection of all the cross products of the form $S_\alpha \times T_\beta$. What is the appropriate index set for this family of sets? Fix some α_0 in Λ; is there another expression for $\cup_{\beta \in \Gamma}(S_{\alpha_0} \times T_\beta)$? What about for $\cup_{\alpha \in \Lambda}(\cup_{\beta \in \Gamma}(S_\alpha \times T_\beta))$? Proofs?

4.66: Suppose $\mathcal{S}$ is a family of sets contained in the real numbers $\mathbf{R}$ with the property that for any $r \in \mathbf{R}$ there is a subfamily $\mathcal{T}$ contained in $\mathcal{S}$ such that $\cap_\mathcal{T} T = \{r\}$. Prove that $\cap_\mathcal{S} S = \emptyset$.

4.67: Give an example of a family of sets indexed by the real numbers $\mathbf{R}$ and which is <u>decreasing</u> (in the sense that if $t < r$ then $S_t \supseteq S_r$ and $S_t \neq S_r$), and which has the property that $\cap_\mathbf{R} S_r = \emptyset$.

4.68: (De Morgan's Laws) Suppose that there is some fixed universal set U and some family $\mathcal{S}$ of subsets of U. Conjecture formulas for $(\cap_\mathcal{S} S)'$ and $(\cup_\mathcal{S})'$ and prove them correct.

4.69: It turns out that any family of sets $\mathcal{S}$ may be viewed as an indexed family of sets in the following way: we will use $\mathcal{S}$ *itself* as the index set, so what is needed is a surjective function from $\mathcal{S}$ to itself. That seems easy enough! This leads to expressions like $\cup_{S \in \mathcal{S}} S_S$, which are a little disturbing psychologically but fine mathematically; each set is indexed by itself. Prove that the union and intersection of $\mathcal{S}$ viewed as merely as a family and as indexed by itself in this way actually coincide.

4.70: Suppose $\mathcal{S}$ is a family of sets. One may define a relation R on $\mathcal{S}$ (see Exercise 4.57 for the definition of relation) by requiring $(S, T) \in R$ if and only if $S \subseteq T$. There are various properties a relation may have such as symmetry, reflexivity, and so on (see Exercise 1.24 and Exercise 1.88 for definitions). Which properties does R have? Proofs?

4.71: Critique your reading of this section and handling of the problems and proofs posed in it. It might be worth holding in mind that in another text's version of this section, there were 10 problems that were really concrete examples of families of sets, indexed families of sets, their unions and intersections, and so on, and 11 in-text concrete examples. While that author and I disagree as to who (author or reader) should provide these examples, we are clearly in full agreement that examples must accompany understanding. I hope you're convinced, since this was my last opportunity.

Appendix A
Theoretical Apologia

The purpose of this brief appendix is to mention, for those who are interested in such things, some of the theoretical underpinnings of the approach, taken in Chapter 1, of reading by example construction. But it should be clear at the outset that this book was not written out of any theoretical perspective. The notes from whence it came were written, as I guess teaching notes often are, out of frustration at my lack of success in teaching certain things. My favorite course as an undergraduate was real analysis; I loved the material and the opportunity for active thought; the concepts were neat and the proofs were fun; I learned a lot. When I now teach the same course, my students struggle and many dislike it; they are unable to move from skills sufficient for calculus to reading independently; they waste lifetimes doing nonsensical things; many simply survive and swear off theoretical material forever. My efforts to bring in Pólya's *How to Solve It* [5] to help things don't seem to work; Schoenfeld's description in the Preface and Chapter 3 (Heuristics) of *Mathematical Problem Solving* [7] captures, better than I can, my hope and subsequent frustration with things that seem so correct and don't work for my students. Through some distillation of experience I came to believe that students who had, and even better, had *constructed*, a concrete example were much less likely to wander in a fog of abstractions doing fruitless things. The notes that resulted, and their refinement over the years, have been until recently embarrassingly free of any theoretical knowledge whatsoever.

I share with many mathematicians a distrust, primarily mere prejudice, of the "education community"; it has seemed upon my few samplings that either its members research and prove with powerful statistical techniques and beyond any reasonable doubt something anybody who teaches knew already, or they speculate philosophically about the unknowable reaches of cognition. Schoenfeld's work and lectures are, at least for me, a productive middle ground. In particular, he observes that Pólya's heuristic strategies actually are whole categories of related strategies, and that attempts by students to implement them from Pólya's general descriptions founder at least in part on this lack of detail (see Ref. [7]). But this observation offers a possibility: isolate, describe in detail, and teach a cluster of strategies for reading subsumed by Pólya's "construct an example." Chapter 1 is my effort to do so.

Only later did I become aware of a body of work that provides language and structure that is at least descriptive of (a model of?) the internal cognitive workings one is trying to affect. A sequence of papers by David Tall, Shlomo Vinner, and others defines and discusses the notions of concept definition and concept image. The concept definition is what a mathematician would call the definition, is the formal object to be considered, and is agreed upon by the community. The concept image is the cloud of things that surround the concept definition in the mind of an individual, including pictures, examples, non-examples, related concepts, previously solved problems, standard techniques, standard problems, useful theorems, informal definitions agreeing with or conflicting with the formal definition, and so on. These authors stress the points that concept images are individual in the extreme, and that we do much of our creative thinking with the concept image and not the concept definition. The paper containing the clearest discussion of these definitions is that by Tall and Vinner [13]; Dreyfus' chapter in Ref. [14] (and the bibliography of the book as a whole) give further information and references.

The research on these issues has concentrated primarily on what happens with students whose concept image of a previously encountered notion (say, limit) creates conflict with the formal definition. Which definitions are elicited in what situations? How do students resolve, or stand not resolving, the contradictions? These questions are interesting and some of their results quite depressing (or challenging, if you prefer).

However, the difficulty of a mathematics student at the junior/senior undergraduate level is, along with the above, the problem of forming concept images at all. Faced with the definition of disconnection in topology, or uniform continuity in analysis, or coset in algebra, my students are, and unfortunately often stay, refreshingly free of any concept image at all. In particular, we have not taught them, nor have they learned, any way to build or enrich their concept images independently; they have been encouraged by years of previous mathematical experience to believe that such enrichment is the role of the teacher. Yet at about the junior/senior level we begin to expect them to read and understand comparatively adult mathematical material (written without the obligatory eight or nine examples per section in traditional calculus texts, for example). Our exhortations for them to "read actively" fail because they have no techniques with which to do so except the highlighting pen. The unhappy results for first theoretical courses are as described in the introductions to the present work.

Does the sort of reading by example described in Chapter 1 work or at least help? My experience says yes, and some very preliminary research suggests a positive answer. As well, there is plenty of anecdotal evidence that students feel better and try harder when they believe they have some effective techniques. But real research ought to be done. One measure of effectiveness makes contact with another body of research concerning how experts and novices classify or sort problems. In brief, experts sort problems into groups according to "deep structure" (for example, the physics principles used to solve them, such as conservation of energy) while novices sort problems into groups based on "surface structure" ("this is a problem about balls") (see Ref. [15]). Retrieval of relevant principles and techniques from the past base of solved problems has been shown to be adversely affected in novices by this sort of coding (see Ref. [16]). An interesting question for research would be whether problems made up *by the student* would be resistant to miscoding because deep structure would have to be understood in order to create the problem in the first place. A caution for all of this is, as Schoenfeld argues powerfully in [7], that issues of heuristics are thoroughly interwoven with issues of "resources" (subject matter knowledge), "control" (executive decision making), and others.

A word may now be said about the latter chapters of the present work. Mathematics is unusual in that when doing proofs one eventually must actually manipulate the concept definition (not merely the concept image) and, further, do so at a relatively high degree of formality. Chapter 2 is all about the mathematical community's in-group coding of those manipulations; Chapter 3 is the most basic manipulations of the formalities. The approach there clearly places me in the "more logic" camp of the long debate over how much logic to include.

In spite of the above remarks, the reader is hereby duly warned that the techniques here are without justification other than that gained through experience. A side interest on my part in mathematics cognition notwithstanding, this book was written by a mathematician and teacher, and can really be evaluated only on the basis of its success in aiding students. The author welcomes communications in this regard.

Appendix B
Hints

Collected here are hints, suggestions, and (rarely) answers for the problems in the book, both in-text ones indicated by icons and "official" exercises at the end of sections or chapters. These two sorts are grouped together for numbering (although divided by chapters) so there is a single list combining the two. The intent behind giving hints, and not answers, is to encourage you to continue work on the problems even after you have consulted what is to be found here; few things are more deadly for your understanding than getting the answer from an external source, and it is even worse (if possible) if you stop work immediately thereafter. Please work diligently on the questions first, and use these hints wisely.

Chapter 1

Section 1.2

1.1. You passed the first hurdle — you tried to pick a function. It is mostly a matter of courage: PICK ONE! Almost any one will do. But a place to start is to list all the functions you know and then ask yourself, "which one would I like to be quizzed on?" Take that one.

1.2. Read the previous hint.

1.3. What can be made more specific in the question *since we are working with a specific example?*

1.4. Using a specific function helps, so why not try something else specific, at least to get started?

1.5. Or suppose $x_1 = -2$, so $x_1^2 = 4$. Need $x_2 = -2$ to get $x_2^2 = 4$, or will something else work?

1.6. Did you plot the pair of points you found above that showed failure of injectivity? If you find another bad pair of points and plot them, can you see a pattern?

1.7. Reread as needed Hints 1.3–1.6.

1.8. Try *writing out* the line in the definition of injective with the specifics of this problem filled in.

1.9. Darn well should have drawn a picture (grumble grumble). But compare with the pictures for x^2 and x^4.

1.10. See, and remember for later, Hint 1.8. Algebra helps, too.

1.11. See Hint 1.10, and engrave the hint in Hint 1.8 inside your eyelids. There's a good general rule: if you use a technique more than twice, it is worth remembering.

Section 1.2.1

1.12. One way to start the set builder business is to take the set you got as a list and describe it some other way. Or think of some of the standard subsets of the integers.

1.13. Worth remembering: your collection of examples is not complete if you didn't use both sets given as a list and sets given by a condition. By the way, the words "union" and "intersection" are not arbitrary but actually help you remember their mathematical meanings if you look at their ordinary meanings. In spite of what you think, many of the mathematical words have this pleasant property if you would only look for it.

1.14. A trick we'll come to later is to use A and B so small (really small!) that you can write out all the ordered pairs, first element from A, second from B. Put set brackets around the list and you have your first example. Also, it may help to make the elements of A look very different from the elements of B, at least to start. Also, near the end of things when you are comfortable, suppose A and B are sets of real numbers, so $A \times B$ is a set of ordinary ordered pairs. What does the picture in the Cartesian plane look like (the graph, if you like) of various $A \times B$? We should mention that the name "Cartesian product" comes from the Latin form of the name Rene Descartes, who studied these for the real numbers as a wonderful aid to visualizing functions.

1.15. If you start with familiar real-valued functions from calculus, you will probably first think of the set S as being the whole real numbers. That's OK as a place to start, but for the sine function, for example, think of some smaller sets containing the range. What is the smallest set containing the range? It is also useful to use some smaller examples (a trick we'll get to later), like a function from $\{1, 2, 3, 4\}$ to $\{a, b, c\}$.

1.16. Some small examples (as in the previous hint) help a lot here.

1.17. A four-point vertex set is enough to get started; you might consider briefly how many "different" graphs there are on a four-point vertex set. A puzzle involving graph theory you may have seen is how to draw six dots in two groups A and B of three each, and then connect each dot in A to each dot in B *without* any edges crossing except at vertices. Give it a brief try, then get back to work. One other thing to note is that the shape of the edges, their lengths, and where the vertices are located are irrelevant for what is being considered here. Finally, a small example is the shortest possible walk — a list of one vertex.

1.18. One way to think of this is as capturing when two graphs are the same except for the labels of the vertices. If you draw a triangle, labeling the vertices A, B, and C; and I draw a triangle labeling the vertices X, Y, and Z, they are not the same graph but they are isomorphic. Question: after a few basic examples, can you make two graphs that are isomorphic but don't appear so? (Hint: you can do so with a set of four vertices, if you deliberately introduce unnecessary crossings of edges.)

1.19. It is worth noticing that the easiest way to get a graph that is not connected is to take any graph, add a vertex to it, but connect that vertex to nothing. Also, to play sufficiently with "cycle," you might have to use a vertex set of six points, say.

1.20. One way to find some examples is to be creative with subsets of $A \times B$. How many are there? Also, one way to come up with some examples is to use the natural meaning of "relation" as a guide. What about the relations "is a parent of" or "is in the same generation as," for example?

1.21. The hint for the previous exercise (which is, after all, more general!) works just as well here.

1.22. Quite precisely, we want the pair (a, b) in our set exactly when $a < b$. So, for example, $(4, 2)$ ought not to belong.

1.23. You ought to be disappointed if you don't find the relations for what you used to call "$\leq$," "$=$," "$>$," and "is an integer multiple of." In more general settings, you could, of course, work with these relations extended to the integers or whole real numbers (of course, listing the whole relation is time consuming — just list a sample). Can you find any relation on the collection of all functions from the real numbers to the real numbers?

1.24. You are now rewarded for doing Exercise 1.23 so thoroughly, as well as all your creativity on Exercise 1.21.

1.25. Probably only one of the relations you have created so far is an equivalence relation. Make sure you identify it as your old friend equality.

For another positive example, consider *congruence* as a relation on triangles. Another way to get equivalence relations is to start with what you have to (reflexivity), throw in another ordered pair, and then throw in what you have to in order to guarantee you have an equivalence relation.

1.26. You are exactly in the situation of the question asked in 1.1, namely you need a function to play with. Clearly it is time to assemble, once and for all, an organized list of all your old friends (and enemies?) from calculus. Play with a few; keep it simple.

1.27. Here you need an organized list of sample subsets of the real numbers. What about collections of numbers you knew before the real numbers? What about intervals? What about making up some bizarre looking sets by combinations of the above?

1.28. Here's some reward for doing Exercise 1.27 well. Pictures (on the number line) may help a lot with this problem.

Section 1.3

1.29. You want a function satisfying the hypotheses. What are the hypotheses?

1.30. No hint available for this problem.

1.31. Why isn't $x^2 \arctan(\log(x))$ as good?

1.32. Does $f(x) = x$ have any properties not generally enjoyed by other functions?

1.33. Well, why not 0 and 1 then?

1.34. Any time you think of a real-valued function you ought to help yourself by drawing it.

1.35. A fair objection to the loose statement of the problem is that there are lots of lines with slope

$$\frac{f(5) - f(2)}{5 - 2} = \frac{25 - 4}{5 - 2} = 7.$$

True, but for most of them the 7 was not obtained by this *particular* division in the usual "change in y over change in x" procedure. Which one is obtained this way?

Section 1.3.1

1.36. If you want to go a little farther, can you find some examples both with sets of three elements and with familiar sets of real numbers? Can you find various examples in which (some or all of) the resulting sets have no elements?

1.37. See the previous hint.

1.38. No hint.

1.39. One way to start is, of course, with $S = \{1, 2, 3, 4\}$ and try to build a very small relation with the properties assumed. Can you keep it from being reflexive? Suppose you assume only symmetric and transitive; then can you do it?

1.40. Two sets is easy. How few sets can you partition A into (yes, one is allowed). How many? Can you find all the partitions of the three-element set $\{1, 2, 3\}$?

1.41. Look at your previous examples. Can you find a pair of partitions of some set, one of which is finer than the other? Can you find a pair *neither* of which is finer than the other? How small would you have to get your sets to be sure your partition is finer than any I might come up with?

1.42. No hint available, just work.

1.43. Make sure you actually have a cycle in your graph! What, for example, is the simplest graph with a cycle, say, on six vertices?

Section 1.4

1.44. You do want an a and b you can compute with; how do you feel about $\sin 2$?

1.45. One way to think of good things to do is go back to what you did for the square function.

Section 1.4.1

1.46. No hints available for nonmathematics items.

1.47. This is harder than it looks, since you have to choose a function, and then an a and b so that the conditions hold. You surely need a function negative some places, positive others. Pick that first, and then a and b. One good example would be to cook up some quadratic function, since you could actually solve for the c you need.

1.48. If a relation is a subset of $S \times S$, this comes down to large and small subsets of $S \times S$. How about the subset with no elements for a small one? For another example, can you think of something you might want to call a "circle" relation?

1.49. The first thing to do is to try this on a nice small relation. For example, suppose $A = \{1, 2\}$ and $B = \{a, b, c\}$, and the relation R is the set $R = \{(1, b), (1, c), (2, a)\}$. Write down the line corresponding to the definition for R^{-1}. For example, is $(b, 2)$ in R^{-1}? Once you understand this, you can use the definition on everything you got from Exercises 1.20, 1.23, and so on. And for Exercise 1.23, do any of the R^{-1} look familiar?

1.50. This is mostly a matter of having a good stock of relations, a reward for doing previous exercises thoroughly or an incentive to do them now.

1.51. You'll help matters by using some small sets; no more than three elements per set will do. Some other examples could come from standard subsets of the integers.

1.52. Try this on your old friend the square function, and for goodness' sakes, draw a picture. Using $t = 1/2$ is a good start. Fixing some x_1 and x_2 and looking at what you get out of all the various t might help too. A glance at Hint 1.35 might help.

1.53. Is your collection of examples of sequences of all kinds in good shape? If not, you are at the same place you were when you needed Hints 1.1 and 1.27; you need to get past examples in an orderly package. One example that might be a little useful would be some constant sequence.

1.54. Start with a very small equivalence relation, say on $T = \{a, b, c, d\}$ again. Suppose a is equivalent to b, and then do anything else you like, and everything you must to ensure you really have an equivalence relation. Be careful with this first, and then you can finally start on the problem.

Section 1.5

1.55. It is OK to start with real-valued functions, but realize that you usually think of them as defined on the whole set of real numbers and also as mapping into the whole set of real numbers. This rather limits the domain and codomain! Start by thinking of functions that can have a codomain smaller than $\mathbf{R}$ (the set of real numbers). Move onto some with a smaller domain.

An entirely different place to start is with functions defined on a very small set, say, $\{1, 2, 3\}$. There's lots of freedom to play with the codomain now, even for "familiar formula" functions.

Section 1.6

1.56. Of course you have chosen some specific function and some specific set B. Take the following hints one at a time.

1. If you aren't getting anywhere, try to make B simpler.

2. Does B still have a lot of points in it? Try a set with very few points.

3. If you are still a little stuck, it is a good idea to make sure that you have, with your specific f and simple B, *written out* the line corresponding to (1.2). Make sure B is a subset of the codomain of the function. Stick with it.

1.57. To make $f^{-1}(B)$ small, just cut down the size of B relentlessly. For the other examples, experiment with a familiar function.

Section 1.6.1

1.58. For example, you surely know one element in $f^{-1}(B)$ if B happens to contain 2. Another advantage of this example is that you can examine $f^{-1}(B)$ for just about all possible B. Do so.

1.59. You may need a new codomain to make f injective or surjective. After you are done, you might note that in making f injective you also made it surjective. How can you modify the example so you can examine these two ideas separately?

1.60. Recall that you came up with some examples of injective functions in Section 1.2. And you could use the noninjective functions you constructed from that section, just to be thorough.

1.61. For a start, shade all the things in $f^{-1}(\{2\})$, where the function f is the one you built for Exercise 1.58.

Section 1.6.2

1.62. Important first step: pick X (small, right?) and fix it, and pick A and fix *it*. That is, juggle things for this particular X and A, but don't try to juggle with them at the same time you are doing other things. Try writing χ_A down as a table of values. Some worthwhile special cases: what if $A = X$? What if A is the empty set? Then move to functions on the real numbers.

1.63. WARNING, WARNING! In our discussion of $f^{-1}(B)$, B was in the codomain. Where is B for this definition? That understood, you may as well reuse the function you first chose for the $f^{-1}(B)$ discussion, at least to start with.

1.64. A look at what you did for Exercise 1.59 might help.

1.65. If it is hard to think of a function (an old friend) as a relation (a new acquaintance) with a special property, look at your recent definitions of some functions as *tables of values*. See the connection? The ordered pairs are almost staring at you. See why there are some relations that are not functions, but all functions are relations?

1.66. As part of a table of values, one might have 2|4. In the function (set of ordered pairs) one might have $\{\ldots,(2,4)\ldots\}$. In the functional notation, one might have $f(2) = 4$. But suppose you had also $(2,5)$ in the set of ordered pairs. What does that translate into in the functional notation? NOT ALLOWED.

1.67. Use small examples and a fair amount of work. That's not much of a hint, but that's what it takes.

1.68. Graphs of five or six points are probably needed to get the full flavor. Observe that if you find a graph in which every vertex has eccentricity one, although it may be a small example for distance, it is a large example for number of edges in a graph.

1.69. It's tempting to skip over this one, but don't. This is where you nail down what you learned for future use. For you athletes, it is like game films, which can be embarrassing but instructive. For you musicians, it is like the tape of the concert, which can be embarrassing but instructive.

Section 1.6.4

1.70. We already encouraged you to do extreme examples when considering characteristic functions with small examples. Did you do it in the case in which $X = \mathbf{R}$, too?

1.71. How big could B be? How small? What if f is surjective, or not?

1.72. Have you noticed something about functions both injective and surjective on small sets, namely the number of elements in their domain and range? But some extreme examples might be found with functions on sets that are not finite, like the real numbers or integers.

1.73. You have probably done most of this one place or another, so use this opportunity to pull it all together and organize it. Alternatively, make up a long list of things you should probably find examples of for these definitions, and then see how many you've already got.

1.74. The first thing to note is that all the elements of the table must be from among e and a, since they are all there is in this particular universe. Then, think of this identity element e as being like 1 for ordinary multiplication (this is more than an analogy: 1 *is* the identity element for the operation of multiplication). If you are asked to fill in a multiplication table, the part where you compute 1 times things is easy, right? For the remainder, you got cancellation for a reason, and you have only two choices for the remaining entry in the table. Eliminate one by getting a contradiction if you use it.

This is rather a funny exercise, since we haven't really given the definition of group. But when you do get it, realize that you already have an example you generated yourself.

1.75. The hints for 1.74 is extremely relevant here. To fill in the rest, you have to try a guess for, say, $b * b$. Careful use of cancellation will allow you to eliminate all but one of the possibilities.

1.76. First pick p! Then pick some k and try to find some of the appropriate n to go in $\mathcal{O}_k$. Or, you can pick p and then simply grab some n and figure out which $\mathcal{O}_k$ they belong in. Once you get the $\mathcal{O}_k$ the rest isn't too bad. For part c), you may have to try various values for p.

1.77. Suppose you had to map a small finite piece of the first injectively onto a small finite piece of the second. What would you do? Also, Exercise 1.72 should be reviewed in light of this example.

1.78. Probably you simply have to organize what you've done before. Make sure you see that a reflexive relation can't be too small.

1.79. We already remarked that relations are just subsets of $A \times B$, and there is one biggest subset and one smallest. But how "big" or "small" can a function be if A has three elements, for example?

1.80. One thing to note is that we allow only one edge between any two vertices (there are things called multigraphs where you allow more, but that's another story).

Section 1.7

1.81. When starting by dropping the first condition, realize that the third assumes it implicitly when "$f(a)$" is written. Therefore it is enough to find a picture that fails condition (1) but has condition (2), since (3) is ruled out on a technicality. As for functions that fail to be defined at a point, you might remember several flavors from calculus: those with a "hole" in an otherwise normal looking function, those with a jump and a hole, and those with vertical asymptotes. After you have done some work yourself, it is perfectly fine to consult your old calculus book.

1.82. The hints for 1.81 may be helpful, changed appropriately.

1.83. You may as well add a good collection of these "piecewise" defined functions to your organized collection of real-valued functions.

1.84. You have to have a value, and you have to have a limit. Given a picture, one is easy to change without much redrawing, and the other is not.

Section 1.7.1

1.85. Of course, you looked at your collection of examples from the previous problem, and if you had both an example and a non-example you might have felt done. Can you find a sequence increasing but not bounded above? Bounded above but not increasing? And produce some more examples; if an increasing sequence has a limit, is it bounded above? (If you have trouble thinking of an increasing sequence with a limit, try some sequence of the form "2 minus a little bit," "2 minus a little bit less," ....) If an increasing sequence is not bounded above, can it have a limit? (If you have trouble producing a sequence not bounded above, think of the sequence of integers we use to count with.)

1.86. Every ordinary arithmetic operation is an example, although it is easy to forget that even the addition of three numbers is defined in terms of repeated addition of two numbers. It's hard to come up with standard examples of operations that involve three inputs, but fairly easy to come up with ones that have only one input. How about taking the negative or absolute value of a number? For that matter, think of any real-valued function. For operations on the set of functions, what about adding two functions, etc.?

1.87. Some good examples to consider are multiplication of the real numbers, the positive integers, the positive fractions, matrix multiplication, and the operation defined on a set S of your choice by $a * b = a$ for all a and b (that is, take the first thing). You might also consider the two (small) groups you constructed in Exercises 1.74 and 1.75.

1.88. All can be done on $S = \{1, 2, 3, 4\}$. Supposing one wants a reflexive relation, there are certain pairs all of which must be present, and then you may throw in anything else you like. To fail to be reflexive, exclude one of the pairs, and throw in anything else you like. To be irreflexive, you must exclude all of these special pairs, and throw in anything else you like. The others are similar.

1.89. One way to produce an interesting non-example is to note zero is not positive. Another way is to ask what happens if all the terms a_n are *negative*. Should this also be called an alternating sequence?

1.90. One thing to think about is how many edges must a graph on n vertices have in order to be connected? Is this number sufficient for *all* graphs on n vertices, or is it simply a minimum?

1.91. This looks intimidatingly technical (at least to me). But suppose you start with a graph with lots of edges? Can't you find one of these perfect pairings? Now start deleting edges — can you still make it work? Also, graphs that have the rough shape of a circle can be done.

1.92. One way to get some examples is to draw some graphs. A way to get some non-examples is to require of a small number of vertices a large number of edges. Can you find some more interesting non- examples? By the way, can you make a conjecture about the sum of a graphic sequence?

Section 1.8

1.93. In some ways a theorem is like a good contract. A contract is a (legal) guarantee that if you fulfill your half of the bargain (the hypotheses), the other party will fulfill its half (the conclusion). What would make you think the contract was broken? The nice thing about mathematics is that if the theorem is correct, this can't happen.

1.94. Well, what would the theorem guarantee if we carelessly choose an f continuous at a and at b and also differentiable on (a, b)?

1.95. One way available in this unhappy world is for you to consult your old calculus text (unless you sold it), since as we have pointed out before most calculus texts, trying to make sure you don't hurt your brain by thinking with it, provide a dozen examples for you. But it's better to do things on your own; recall that the derivative at a point gives the (unique) slope of the tangent line to the curve at the point. Therefore, if the curve appears not to have a tangent line at all, or appears to have "several," you probably have a point where the function has no derivative. Also we remarked earlier that if a function has a derivative at a point then it is continuous at that point; therefore (by the contrapositive, actually), functions not continuous at a point are not differentiable at that point. Here are some more examples. Finally, you may remember the words "corner" and "cusp" from elementary calculus, which give some nice non-examples.

1.96. A hint that should be unnecessary: insert a and b and the line passing through $(a, f(a))$ and $(b, f(b))$. Is there a point c of the type required?

1.97. One of your examples probably does this, and the one hinted at surely does.

1.98. The nice thing is that all those nondifferentiable functions you built for 1.95 by making them discontinuous may be used here, at least if you move a or b to the appropriate place.

Section 1.8.1

1.99. There is a positivity condition, a negativity condition, and a continuity condition. If you have one but not both of the first two, it is easy to come up with an example (even a continuous one) violating the conclusion by just staying the heck away from the x axis. If the continuity condition is all you get to violate, take a continuous example (which will therefore *satisfy* the conclusion) and destroy continuity at the crucial point c.

1.100. There are two conditions, so for a non- example you can violate both, just the first, or just the second. Small examples suffice!

1.101. Start with violating continuity and use, essentially, the trick at the end of Hint 1.99. For continuous examples with a set other than a closed interval, try taking a simple function (a really simple function) on some closed interval and then, by removing two points, turning the *closed* interval into another familiar set. What happened to the maximum point? Alternatively, the function f given by $f(x) = 1/x$ is worth considering.

1.102. If you keep the condition "increasing," you did a lot of useful exploration of violations of the "bounded above" condition in Exercise 1.85. For violations of the increasing part, while keeping bounded above, realize that nobody said you had to be bounded below. Or, construct some sequence that wobbles back and forth between 0 and 1.

1.103. Try reading these hints *one at a time*.

1. Functions on small sets, of course.

2. Make the elements of the three relevant sets look quite different (a, 2, X).

3. Start with f and g injective, and see if $g \circ f$ is.

4. Take your previous example, and (adding a point to the domain of g if necessary but not touching f) make g do something destroying injectivity of g which is irrelevant to $g \circ f$. Is $g \circ f$ still injective?

5. Start with an example in which f fails to be injective. Is there anything you can do with g to save $g \circ f$? A two-point set for the domain of f is quite enough.

1.104. Start with a small set A and build the relation by hand on as much of some set B as you need. You may need to throw away some of B to ensure R is surjective. Then just compute R^{-1} and note that in many ways it looks rather like R.

1.105. Luckily, there is only one hypothesis on the walk.

Section 1.8.2

1.106. If you are having trouble, in that all of your non-examples do fail the conclusion, take one of them and make it "wiggle" more between a and b.

Section 1.8.3

1.107. One way to produce examples is to start out by giving yourself c (that is, a place of crossing the x axis) right away. Then, go crazy — destroy continuity, destroy positivity or negativity conditions, do anything you like.

1.108. The hint for Exercise 1.107 gives the right idea, suitably changed.

1.109. Suppose you take a sequence of the type guaranteed to converge by the theorem, and change the second term. What happens to the convergence? Can you violate the hypothesis of the theorem by so doing?

1.110. What would it mean if you were unable to find an example in which the conclusion held with the hypotheses violated?

1.111. Draw a picture (put points on the real number line) of an increasing sequence with a limit; does it appear to be bounded above? What would be an upper bound?

1.112. The question is, of course, whether f injective implies $g \circ f$ injective *no matter what g is*. The g to challenge this would be a g as far from injective as possible. Try a small example with a really disgustingly non-injective g.

1.113. Note that it is possible to have a repeated vertex without a repeated edge. See why you have to ask?

Section 1.9.1

1.114. Try the following hints, one at a time.

1. One way to start is to come up with a sample f and h that do the right thing, and worry about some g later. If you can find f and h, even just with a picture, sketch some g "between them" in the sense of the second hypothesis.

2. Notice that the second hypothesis gives information about how f and h compare (ignoring g) point by point. How would this show up in a picture?

3. What do you want f and h to do? If they were continuous, so their limit at a is their value at a, they would have to come together at a point, right? And, as noted in the second of these hints, f and h would be in a definite relationship elsewhere.

4. An example of f and h is to use f the square function, and h the negative of the square function. Draw in g to be anything you like. Don't stop with this example, though.

For the final question, ask yourself what would change if you put "holes" in the graphs at the appropriate a. Does the limit of f or h change? Does the limit for g change?

1.115. Of course, you are using a nice small set. You need some subsets of it whose union is the whole set, and which are themselves disjoint (that is, have intersection the null set). A good non-example would be to come up with some disjoint sets whose union is not the whole set; another would be some sets whose union is the whole set, but which are not themselves disjoint. If your original set had, say, four elements, what is the largest number of sets you could use? What is the smallest?

After you are comfortable with the definition, you need to start over, with an equivalence relation, to make sense of this E_x stuff. Do so; a good equivalence relation might be one on a set with four elements.

Finally, realize that the statement of the theorem has to be read carefully. It might be that E_x and E_y overlap somewhat, but Exercise 1.54 tells what happens in this case. For forming the partition, then, this set only counts once.

Chapter 2

Section 2.1.1

2.1. Here's a place to start. Number 1 must have been something of a shock, since you have no idea what A and D are; *it* can't be first. Indeed, you have a right to expect that some sentence supposed to come before this one tells you. This lets you pick out a couple of good early sentences. Likewise, any sentence assuming something given in the hypothesis of the theorem is likely to be an early one. Finally, something giving the conclusion is likely to be an ending one.

2.2. Find the first sentence of the proof, by elimination if necessary. This gives a two-step form for the proof. Can you group sentences by, at least, which of the two steps they are part of?

2.3. Besides the usual first and last clues, there is a key word in Number 4: also.

2.4. We hope you remember something about induction, namely that it is a standardized two-step process. Other than that, the signal "three equations" in Number 4 is a good deal of help.

2.5. Finding the first sentence is at least as important as in previous exercises. Also, the appearance of "t_n" may have been something of a surprise, but finding the right sentence will not only explain it, it will give a valuable clue to the structure of the proof as a whole (especially if the first sentence didn't quite give enough).

Section 2.1.2

2.6. The temptation will be to think a little bit, perhaps, and then skip down to where you expect to find our list. A better investment will yield a great deal more.

Section 2.1.3

2.7. No further hints available.

2.8. Ditto.

2.9. There may have been several violations, but the gravest is that the reader wasn't told in the beginning what the structure of the proof would be (compare Item 6 in Exercise 2.2). It's likely the one easier to untangle is the one you could have done yourself, namely 2.1.

Section 2.1.4

2.10. See the hint for 2.6.

2.11. Little things are deadly. The "thus" in the third line leads you to expect the third line came from the second. How could it? "And" would indicate better the source of this one. And the "therefore" in the fourth makes you think it came from the third, as opposed to which *two* previous lines?

Section 2.1.5

2.12. Observe that the failure to set the notation makes this almost impossible. The vertices of the rhombus are A, B, C, and D in counterclockwise order, with E the intersection of the diagonals. Now try again to unscramble this before reading further. Note that this is still a mess. One particularly brutal omission of a cue is a "similarly" before the conclusion $DE \cong BE$. And you were wondering how this came from the arguments above. Answer: it didn't.

2.13. Again, the question to be asked is not whether you can do the proof, but whether you are in fact having to do/reconstruct the proof you are supposed to be reading. Insert the cues to make this unnecessary.

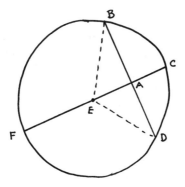

Section 2.3.1

2.14. Individualized hints are rather difficult, but you might try geometry.

2.15. Ditto.

Section 2.3.2

2.16. Ditto.

Section 2.3.3

2.17. This one isn't too hard (note that you never do assume that h is injective), but how well is it cued?

2.18. This one is much trickier than it looks. First, you probably were looking for a proof form for implication. Fine, except, what is the implication being proved? After some work you may finally hit on "If G is a group, then its identity element is unique," which is correct but much less obvious than in the previous problems.

Second, uniqueness proofs have a structure all their own. The direct way is to let a and b be names of the thing with the desired property (note that we do not assume there are two of them for a contradiction, just that there are two names, just as 2 and $4/2$ name the same thing). We then show that these two names name the same thing, that is, $a = b$. Thus there really is only one *object* with the property.

Finally, note that if you are cheating a bit and doing these by process of elimination ("no contradiction," OK, did we assume "not something," No? OK, must be direct proof ...) you may have gotten the right answer for the wrong reasons.

2.19. Compare with 2.18, and note that the insertion of the single word "distinct" changes this to a proof by contradiction all by itself. Now we *do* assume that there are two things with the desired property, not just two names that might name the same thing, and get a contradiction. Note also that this could not be done as a proof by contraposition, because the contradiction we get is not a contradiction of our hypothesis (G is a group) but of something else.

2.20. This one is easy. A better question than "what is the form?" might be, "in how many places is the form cued?" Note that in all other ways the proof is completely opaque, but at least you understand its large-scale form, and that's something.

2.21. This is what you might call a direct construction; we need something, and we build it. We'll return to these in a later chapter. However, the cuing is not great: when did you know that $-b$ was going to turn out to be the lower bound? Where could/should you have known?

2.22. See the hint for 2.21.

2.23. Make sure you decide about, not only the main proof, but the subproof(s).

2.24. No hint available.

2.25. One question to answer here is, how many cases were eliminated by "W.L.O.G."? Put differently, how many different lists of nine things might be required to put them in decreasing order? Whew! Also, note the one-sentence proof by contradiction.

2.26. What straightforward cuing!

2.27. This is a direct proof in a sense, but note that really instead of proving '$A \Rightarrow B$' directly, one announces that everybody knows that '$C \Rightarrow B$,' and starts in on a proof of '$A \Rightarrow C$.'

2.28. Note that this is a direct proof, in the sense that we verify needed properties and call it a day.

2.29. No hint.

Section 2.3.4

2.30. The most efficient might be to use a proof by contradiction, and then see at the end whether a proof by contraposition is really what you did.

Section 2.4.5

2.31. By some sort of elimination, this is a direct proof overall, with the clue in the last line that there are two cases. Drawing a picture (you might know these as Venn diagrams) might help you see the cases.

2.32. This proof has three steps (and when did you find that out? poor cuing!), since there are three things to verify. But the last step has what as its pieces?

2.33. When did you find out what kind of proof this was? The proof itself is rather clever, but the cuing of the two steps is needlessly unclear. Also, what "flavor" of this sort of proof is it?

2.34. How many cases? What do you think of their cuing?

2.35. This one is tricky. About the time you see "holds for $k + 1$," you may suspect something. Also, realize that in spite of the word "case" there are no cases, but in that sentence there is a subproof. What kind?

2.36. There are no cases at all, nor is there induction, and the proof is direct. This problem was really inserted to make you nervous. How much of the proof seems to fit into what we've done so far? Not much. Hold that thought until Section 2.5.

2.37. No cases or induction.

Chapter 3

Section 3.2

3.1. We hope these are familiar, but to fill one in, for each row just deduce from the truth table for 'or' what the entry should be. In general, the result for 'A or B' with A true and B false is what? Same here.

3.2. This is tedious, but not hard once you grasp the system, and everybody ought to do it once. Here is what must be filled in for the first expression:

P	Q	R	(Q or R)	P and (Q or R)
T	T	T		
T	T	F		
T	F	T		
T	F	F		
F	T	T		
F	T	F		
F	F	T		
F	F	F		

Each missing entry is filled in using one of the basic truth tables on the appropriate values of the inputs; so, for example, the entry for (Q or R) in the fourth row is F.

Once you have done this, you may do a similar one for the second expression and compare the final columns. If they are the same, so the same combination of T and F inputs always gives the same result, they are equivalent. Notice along the way the device for making sure all combinations of T/F inputs occur.

3.3. You might look back at some previous proofs by cases in the exercises.

3.4. A glance at Exercise 2.31 might help.

Section 3.2.1

3.5. For proof by contraposition of '$A \Rightarrow B$' you assume 'not B,' work to deduce 'not A,' and then are allowed to deduce '$A \Rightarrow B$.' That replaces things in the template for the direct proof of '$A \Rightarrow B$' quite nicely. Proof by contradiction, recall, has two assumptions, and you may deduce any contradiction. How about, 'C and not C'?

3.6. It probably helps to set up a table like this, using columns as needed, and just comparing the last two.

P	$\neg P$	Q	$\neg Q$	$P \Rightarrow Q$	$(\neg P) \Rightarrow (\neg Q)$
T	F	T	F		
T	F	F	T		
F	T	T	F		
F	T	F	T		

Note that the trick on P and Q to get all the inputs is still being used.

3.7. Tedious, but not hard.

3.8. One you may have forgotten is the deduction of 'P' from 'P and Q.' Also, what would you need to get *to* 'P or Q'?

Section 3.3.1

3.9. Can you build the condition for the set $\{2, 4, 6, \ldots\}$, for example? Or for the set $\{2, 3, 5, 7, 11, \ldots\}$?

3.10. How many variables do you need? What classes of objects do these variables range over?

Section 3.3.2

3.11. Observe that "the graph" is a pronoun and calls for a statement form; "the complete graph on four vertices" sounds like (and is) a specific graph and so requires a constant. Note that this is one way to turn the statement form into a statement.

3.12. Observe that neither "the group" nor "G" ties us down to anything specific. Incidentally, "Abelian" means that the group operation is commutative, and the name comes from the mathematician Abel.

3.13. No hint required.

3.14. "If–then" is a familiar form, so it comes down to the conditions. The simplest approach is to use C for the condition "the square function is continuous at 2"; the problem is, of course, that this is not flexible, since there are no spots for variables like different functions. There is a choice, which is to assume that 2 will always be the point in question, or not. Suppose it is, so the only variable is the function. What do you get? Now suppose that we may sometime want to study functions continuous or differentiable at 3; how can you make a form to prepare us for this?

3.15. Nothing new, except to note that this is true (!). To see why, remember the truth table for '$\Rightarrow$,' and note that since the conclusion is true, the whole implication is true. However, this is not a safe statement to count on if applied to other functions.

3.16. The conditions in hand, this is a simple 'and.' Sorry about that. But the point is that once you have the conditions lined up, things like implication, and, or, not, and so forth, are handled just the way you would before you ever heard of quantifiers.

3.17. Something like $I(R)$ and $F(R)$ might work, with I and F standing for the obvious conditions. Note that a temptation to write $I(F(R))$ must be firmly crushed; that says, read carefully, that "R is a function" is injective (that is, the <u>statement</u> <u>form</u> "R is a function" is injective). But a statement form can't be injective, since a statement form is not a relation.

Section 3.3.3

3.18. It frequently helps to read '(' as "such that." In the second one, you get better English if you move the "such that" after your first effort at translation.

3.19. It may be easier to compare with classmates if you all use the condition "It is raining today" with the variable "days."

3.20. Just keep going.

Section 3.3.4

3.21. See Hint 3.18 and use it in the other direction. There isn't really need to make up a condition for $c^2 = 17$, since the equation really is a statement form, but you can if you want to.

3.22. Who knows what f is? But since it is around, $f'(c) = 0$ is again a statement form, so the hint for 3.21 applies. Please note, though, that what you get at the end of this is still a statement form and not a statement, because of the f.

3.23. The symbol $\in$ from set theory is useful.

3.24. Try these hints one at a time.

1. Note that c really has to do two separate things.

2. Suppose it read "There exists c such that c is in $(0, 1)$ and $f'(c) = 0$"?

3.25. Get the "for all x ..." part first. Then you could insert some appropriate subexpression in the thing you got for 3.24.

3.26. Hint well worth remembering: in this context, "has an" means "there exists an." Some form like $I(G, x)$ for "x is an identity element for G" is useful.

3.27. If "has a" means "there exists a," what does "has two" mean? But there is no special symbol for "there exist two," so $\exists v_1, v_2$ is what's needed.

And how can you make sure they are really *distinct*, and not just different names for the same thing? Different names for the same thing would be equal, but ... . One more, quite common, way to say this is to say "there exist distinct ... ."

3.28. No hint available.

3.29. This is tricky because the word "continuous" is used in two separate ways. One has to do with the continuity of a function at a point (see your work on Exercise 3.14). The other has to do with the continuity of a function on a whole set, which is a different condition and so requires a different notation. After this is sorted out, remember that we have notation for the specific set we care about, namely the domain of the function. For the second one, "each" is the key word.

3.30. You should do these, no matter how else you can do them, by converting them into logical language and negating what you get, translating back into English thereafter. Even stating these correctly, before negation, isn't too easy. In particular, to say that there is no inverse is to say that an inverse does not exist. Equivalent forms for some of the logical statements may help the negations (see Exercise 3.7).

3.31. The likely errors for a), b), and c) are that, in the excitement of dealing with quantifiers, you have forgotten the results of Exercise 3.7 about what is equivalent to the negation of an 'or' statement, and what is equivalent to the negation of an 'and' statement. In particular, the negation of 'P or Q' is *not* anything with an 'or' in it, but with an 'and'; no matter how complicated the things inserted for 'P' and 'Q,' this still holds.

The likely errors for the rest have to do with a tremendous desire on the part of students to give the negation of '$P \Rightarrow Q$' as something with an implies in it. Unfortunately, again as in Exercise 3.7, '$\neg(P \Rightarrow Q)$' is equivalent to 'P and $\neg Q$,' and not to anything else very useful. So after getting past the first quantifier in d) and following, you are faced with exactly this situation, and should not have ended up with any '$\Rightarrow$' signs left over.

PLEASE REMEMBER THIS! It will save you a lot of pain. No matter how much you want it to be, '$\neg(P \Rightarrow Q)$' is not, for example, '$P \Rightarrow \neg Q$.' The statement '$\neg(P \Rightarrow Q)$' is equivalent to 'P and $\neg Q$.'

3.32. Work from the outside in; when you come to something not at the quantified level (an 'and,' 'or,' or '$\Rightarrow$') stop, take a breath, and use Exercise 3.7.

Section 3.3.5

3.33. No hint available except that it is somewhere in Chapter 1.

3.34. See the text following the question.

3.35. This is a bit tricky, but Exercises 3.22 and following have done a lot of the pieces. Also, remember again that the equation is just as good a condition as $E(x, y, z, g)$. Some people find it easier one way, some the other.

3.36. Note that $a < b$ is already a condition, but you may make up some notation for it (two variables, right?) if it helps.

3.37. No more hints available, but keep at it.

3.38. If you do this, how many things are part of $H(f, a, b)$?

Section 3.3.6

3.39. Two hypotheses, one conclusion. Note that you may make up notation for "contained in," but $S \subseteq \mathbf{R}$ is a perfectly good condition (statement form). Also, "closed" is not the same as "not open" (sorry — one is sometimes told that a set is not like a door, because it can be neither closed nor open).

3.40. This is a little nastier than it looks; one really ought to use $O(G)$ to denote the order of the group and something like $P(n)$ as notation for "n is prime," since "prime order" really does mean the order is a prime number. It would be OK to try it first using something like "$PO(G)$" to say G has prime order, but then try the other way. The moral is that since we already know what prime means, we ought to use it, as opposed to making up new notation including prime in it. For a similar example, a function can be something called "even." If we want to consider even and continuous functions, we'd use "$C(f)$ and $E(f)$" (or something), not make up a new animal $EC(f)$. See?

3.41. How about '$C(\{x_n\}, L)$' and '$CA(\{x_n\})$'? Note that what you get, unless you twist things to quantify L, is still only a statement form. Oh, and by the way ... there surely is quantification on $\{x_n\}$, right? It is hard to say this neatly, since $\{x_n\}_{n=1}^{\infty}$ hardly looks like a single name. But it is, the name of the sequence.

3.42. Try these one at a time.

1. "Convergent to some limit" means some limit exists.

2. Suppose it read " ..., then there exists a limit L to which $\{x_n\}_{n=1}^{\infty}$ converges."

The notation of 3.41 may help.

3.43. Note that the "there exists c ..." can practically be stolen from the statement of the Mean Value Theorem. Of course, you recognize that this is the Intermediate Value Theorem.

3.44. Some of this is familiar; if you get stuck, try dealing with the conclusion assuming you already have c in place. Of course, you recognize this as the Maximum Theorem.

3.45. Surely this is about all collections $\{v_1, \ldots, v_n\}$. It may help you to give some name, like V, to the original collection of vectors and some name like V' to the new collection (it makes things easier if you don't have to grapple with such a long name as $\{v_1, \ldots, v_n\}$ while trying to figure things out).

3.46. To say that there is more than one vertex is to say that at least two things exist; see Exercise 3.27 and its Hint.

3.47. Don't make "nontrivial" a condition on a graph, just buckle down and say that there are two distinct vertices. See the discussion in the Hint for Exercise 3.27.

3.48. It might help to have a symbol for the number of elements in a set; make one up.

Section 3.3.7

3.49. No hint available.

3.50. Recall that a function f is continuous on a set S (say, $C(f, S)$) if f is continuous at each point of S (say, $\forall x (x \in S \Rightarrow C(f, x))$). Thus continuity on a set really has a universal quantifier hidden in it. To each x in $[a, b]$ we may apply the universal to deduce something. Similarly for each x in (a, b). If some x happens to be in both, we get the deductions resulting from each of the universals.

3.51. We played with this before, in Exercise 3.14 and following, but surely you remember the truth.

Section 3.3.8

3.52. No hint available, but while your geometry is rusty, it can't be this rusty. Even a good attempt is enough.

Section 3.3.9

3.53. The words "there exist" point to one use of E.I. But the definition of injective is "f is injective if, for every x_1 and x_2, $f(x_1) = f(x_2)$ implies $x_1 = x_2$." This is not an existence claim, you say? Then first negate it using some rules about how quantifiers negate (see around Exercise 3.19). Second, realize that the negation of $A \Rightarrow B$ is 'A and not B.' See where "distinct" comes from?

Also, why does $h(x_1) = h(x_2)$ and $x_1 \neq x_2$ imply h is not injective? Well, it means that there exist x_1 and x_2 so that $h(x_1) = h(x_2)$ and $x_1 \neq x_2$. You will find that this is the negation of the definition of injective (quantifiers and all), this time for h.

3.54. This looks reasonably harmless at the quantifier level, and it is ...almost. Recall that the definition of f an identity for G is that, for all a, $a * f = f * a = a$. So the first equation really results from the application of this universal with a set to e (that is, U.I. applied to the particular element e). The second equation is similar, but results from using U.I. on a definition involving e as applied to the particular element f. This sort of U.I. is so common that it is often completely uncued.

3.55. See the hint for 3.54.

3.56. There are two obvious places, one indicated by "there exists a disconnection" and one by "for each α." The first is a deduction from the definition of disconnection: A set is disconnected if there exists a disconnection, which is a pair of sets (doing something). Therefore, since we assume (for a contradiction, but that isn't the point this time around) there is a disconnection, we may give the pair of sets forming the disconnection the names C and D using E.I. (By our convention this might be C_* and D_*.) We then work with C and D.

The second obvious place is actually a brief, rather uncued, proof of something "for all α" (that is, a proof by U.G.). A longer version might be "Let α_0 be arbitrary. Since p is in the intersection of all the A_α, it is in A_{α_0}. It follows that since A_{α_0} is connected, A_{α_0} is contained entirely in C (take this part on faith — it has nothing to do with quantifiers). Since α_0 was arbitrary, this holds for each α."

The bad news is that there are other places. What does it mean to say that some sets have nonempty intersection? Answer: there *exists* a point in the intersection. So the "p" produced in the first line is actually via E.I. And there is another place where a proof by U.G. is hidden or implied, which is that if for all α, $A_\alpha \subseteq C$ then $\cap_\alpha A_\alpha \subseteq C$. (Alternatively, perhaps this is a previous lemma we don't need to do here.)

Moral of the story: As you said the first time around on this exercise, it is a proof by contradiction. But most of what was going on had nothing to do with that, but with quantifiers.

3.57. Something which helps is to be given a formal definition of bounded above, so here one is: A set S is <u>bounded</u> <u>above</u> if there exists an upper bound for S. An upper bound for S is a number L such that for all $x \in S$, $L \geq x$. The definitions of bounded below and lower bound should pose no problems.

To say that $-S$ is bounded below is to say that there *exists* a lower bound for $-S$, so the proof is really a construction of that lower bound (so the proof is what form?). Our candidate is $-b$; note that b came from (what form?) since S is bounded above. Also, to be a lower bound, one must show that *for all* y (hint, hint) one has $-b \leq y$. And finally, of course, there is the usual "harmless" universal quantification on the set S.

3.58. Here's one start: what kind of quantification would you expect in the definitions of "linearly dependent" and "span"? Also, what is the overall form of the proof, at the quantification level? Read the second sentence of the statement carefully. And, of course, harmless quantification, but on how many objects?

3.59. Where did h come from? Once we show that h is what's needed, what quantifier step comes next? Note that this is the purest form of existential use and proof (that is, the deduction form for Existentially quantified statements and the proof form for Existentially quantified statements); the h from E.I. is exactly what is needed to be the h for E.G. Don't count on it in general!

3.60. No hints available, but just for completeness it is a property of groups that inverse elements exist, which is where b comes from.

3.61. Note that except for the harmless universal quantification on the a_i, this proof is *without* quantification.

3.62. With a clearly quantified definition of injective at hand, the role of x_1 and x_2 will be clearer. Here's a more subtle point: exactly what inference scheme, applied to injectivity of which function and *what points*, allows one to deduce that $f(x_1) = f(x_2)$?

3.63. Of course, there is a quantified inference scheme for this stuff about a and b. There is another scheme going on, though, and it is easy to miss because it seldom gets written out in detail (the results are so self-evident it doesn't seem worth it). Consider how one shows T is nonempty. Since T is the intersection of the sets in $\mathcal{S}$, to show something (in particular, e the identity element of G) is in T one must show it is in each (hint, hint) $S \in \mathcal{S}$. So let S_0 be arbitrary. Since S_0 is a subgroup one has $e \in S_0$. So since S_0 was arbitrary, $e \in S$ for all $S \in \mathcal{S}$. This seems awfully formal to deduce what you surely believed in the third sentence, but it really is there, a whole hidden U.I. proof.

3.64. Note that there are three existence proofs in there, after the universal quantification on a is dealt with. Note that for the first, we don't use E.I. to get a candidate, but actually construct the (quite trivial) walk. But where do the other two candidates come from?

3.65. One might suspect that x and y are arbitrary. But note that here, as you were warned previously, all of the structure for the U.G. argument based on the definition of injectivity $(\forall x \forall y (f(x) = f(y) \Rightarrow x = y))$ is hidden. No cues (like "arbitrary", for example) are given for what x and y are, because the writer assumed that you would know that they had to be chosen and what they were.

By the way, did you spot the use of E.I.?

3.66. It might help to note that the definition of $E \subseteq F$ is $\forall x (x \in E \Rightarrow x \in F)$. Here's a nasty question, though: does the proof by cases come "inside" or "outside" the argument to deal with the above definition?

3.67. Note that the definition of d_b really should have had universal quantification on x and y. Without the definition of metric in front of you, this one is hard, so here it is:

Definition *A function d from a set $M \times M$ to the real numbers is a metric on M if*

1. *$d(x, y) \geq 0$ for all x, y in M, and $d(x, y) = 0$ if and only if $x = y$;*

2. *$d(x, y) = d(y, x)$ for all x, y in M; and,*

3. *$d(x, y) \leq d(x, z) + d(z, y)$ for all x, y, z in M.*

Remark: an "arbitrary" or two to cue the use of U.G. would have been nice.

Challenge problem: where is there a use of U.I. in this proof? Hint: if you rewrite this following the subscripting convention for variables involved in a U.G. proof, you will have a better chance of spotting it.

3.68. If you are tempted to think that there is no quantification here, we tell you sadly that there is lots of hidden quantification in any proof by induction. For a hint of this, n is certainly something or other, probably arbitrary. What is the universal it is involved in the proof of? For more on this, see Section 3.4.3.

3.69. We rewrite the proof following the subscripting conventions so that some things become more obvious. While we are at it, we'll fill in all the quantification and deduction forms, just this once. Prove that for any x, $|-x| = |x|$.

Proof. Let x_0 be an arbitrary real number. There are three cases.
Case I. If $x_0 = 0$ the result is trivial.
Case II. If $x_0 > 0$, then $-x_0 < 0$, so by definition (*U.I. applied to the definition of $|x|$ and the particular point $-x_0$*) $|-x_0| = -(-x_0)$. Of course, $-(-x_0) = x_0$ from algebra. And $x_0 = |x_0|$ from application of U.I. to the definition of $|x|$ and the particular point x_0. From these three equalities we have $|-x_0| = |x_0|$, as desired.
Case III. If $x_0 < 0$, then $-x_0 > 0$, and $|-x_0| = -x_0 = |x_0|$ again by application of U.I. to the definition of $|x|$ and applied to the particular point $-x_0$ to get the first equality, and by application of U.I. to the definition of $|x|$ and applied to the particular point x_0 to get the second equality.

Thus in each case we have $|-x_0| = |x_0|$. Since x_0 was arbitrary, by application of U.G. we have the result for all x.

What's the point? Without the subscripting convention, so instead of x we have x_0, you'll never spot all the various uses of U.I. The reason is that if everything in sight is x, it is hard to realize that the x in the definition of $|x|$ is not the same x that is running around your proof, which is a fixed but arbitrary x. You have a slightly better chance of seeing where you apply things to $-x$, but not much. As you might expect, the results are so sensible and you have to do this so often in proofs that it is almost never cued or written. But you should know it is there. *Any* time you deduce something about a variable you are probably using a U.I.

By the way, this proof, well understood, is helpful for answers to questions in Hints 3.66 and 3.67.

3.70. Again this is an induction proof; note that we started with $n = 2$. Then switching to k, we did the "induction" step; we warn you again that there is hidden quantification here. Try to see which universal is being proved via this arbitrary "k."

3.71. This is perhaps the hardest proof in the lot, but it is also the best cued. There is a definition with two universal quantifiers and whose result is an existence claim. You want to prove something fits the definition, so you use U.G. as the form to deal with the universals and then produce the thing you need to exist, so using E.G. Along the way, you use the definition twice with the U.I. form; here the key is "applied to," and then having applied it, you get to use E.I. to produce something.

Why you might choose to apply U.I. and the definition to these particular things is a much different question. That would be discovering the proof; right now, be content with following the tracks.

3.72. This one is fairly harmless. If the subscripting conventions were being followed, it would be easier to spot the fact the "associativity" really uses a U.I. to be applied to the particular things of interest. And there is the standard universal quantification at the front, which the proof deals with by ignoring it completely! (But the U.G. form is still there.)

Section 3.4.1

3.73. It is worth remembering that, given the statement forms needed, things about 'and,' 'or,' 'not,' 'implies,' and so on are handled for quantified or unquantified statement forms just the way they are for P, Q, and so on.

3.74. Please use the subscripting conventions even if you don't really approve of them. There is some "harmless" universal quantification to clear out of the way first by the usual use of U.G., so there's the beginning and ending of the proof. After that is done, you ought to be facing a certain set containment to prove. Probably the definition of containment is needed; we used it in the proof in Section 3.4.1, but here it is formally:

Definition *If A and B are sets, then $A \subseteq B$ if and only if $\forall x(x \in A \Rightarrow x \in B)$.*

Now, of course, you see both a U.G. and an implication to be dealt with, so do so. As you go on, you will have to deal with the definitions of both union and intersection. The desire for a "union" conclusion might recall to you argument by cases, since union is essentially an 'or' condition. Remember to work backward, too!

3.75. How many things need to be universally quantified to turn the equation (a statement form) into a statement? There's one subtle point: the statement is an implication, since there really is a hypothesis (that A

and B are subsets of the domain of f). And then the first step and last step of the proof come from escaping those universals. What about the second and second to last step?

3.76. Look at what follows the "thus" and think of what to do.

3.77. Try writing it down without peeking, quantification and all. If you have to peek, it is in a previous, and fairly recent, hint.

3.78. More universals to prove.

3.79. See a hint near the end of Hint 3.74.

3.80. If you haven't found the definition of $f(A)$ already, you cheated by not reviewing the definitions in the first place. The one you need is in Exercise 1.63. Grrrrr.

3.81. This is the oranges–apples problem. In this case it is more like oranges and cows.

3.82. We want $f(z) = y_0$. Well, it is possible that $f(y_0) = y_0$, but this doesn't happen too often. The point y_0 is known to be in the range of f, and we are looking for a point in the domain. There are plenty of functions with domain completely different from range. Nope, not y_0.

3.83. No hints available.

3.84. No hints again. Try hard, though, because nobody wants to do this often, but everybody ought to do it once or twice.

Section 3.4.2

3.85. You will surely need to have assembled the definitions of union, intersection, and set containment. It probably would help to draw a picture. BUT the most important thing is to start from the conclusion and work backward. Don't look at the next lines until you have tried again, but if you don't have them *before* you start working with definitions and the hypothesis you are probably astray. You should have as last lines (and with the associated lines at the beginning of the proof in place):

> ...
> Thus $x_0 \in (A_0 \cap B_0) \cup (A_0 \cap C_0)$ by(?).
> Thus $x_0 \in A_0 \cap (B_0 \cup C_0) \Rightarrow x_0 \in (A_0 \cap B_0) \cup (A_0 \cap C_0)$
> by direct proof.
> So $\forall x(x \in A_0 \cap (B_0 \cup C_0) \Rightarrow x \in (A_0 \cap B_0) \cup (A_0 \cap C_0))$ by U.G.
> Since this is the definition of set containment, we do have
> $A_0 \cap (B_0 \cup C_0) \subseteq (A_0 \cap B_0) \cup (A_0 \cap C_0)$.
> Since A_0, B_0, and C_0 were arbitrary, the result holds in
> general by U.G.   ■

The final hint is that you will need to use a proof by cases.

3.86. The hints for the previous exercise, suitably modified, work fine. Make sure you get the concluding lines first, as far up as you can! Did you draw a picture?

3.87. All you need to throw in here is the definition of set complementation. But do draw some pictures.

3.88. You are in great need of the definition of injective. Remember to work from the conclusion end! You might be helped by drawing some pictures. After you have tried those hints, here is another: you ought to get the ending lines below (don't peek before you've tried):

$\quad \cdots$
Thus $x_1^0 = x_2^0$ by (?).
Thus $g_0(x_1^0) = g_0(x_2^0) \Rightarrow x_1^0 = x_2^0$ by direct proof.
So $\forall x_1, x_2 (g_0(x_1) = g_0(x_2) \Rightarrow x_1 = x_2)$ by U.G.
Since this is the definition of injective, applied to g_0,
we have g_0 injective.
Since g_0 was arbitrary, we are done by U.G.   ■

3.89. Along with the hints above, realize that the definition of surjective involves showing something exists. Thus the overall form of the proof is a grand E.G. Doing a few numerical examples might help, as might a picture.

3.90. Work backward from the conclusion desired for g (or, better, g_0), inserting definitions as you go, before you try to do anything with the injectivity of f_0. Well into the middle of things you ought to have assumed $g_0(x_1^0) = g_0(x_2^0)$ and be hoping to deduce $x_1^0 = x_2^0$. In order to do this, you may finally use the definition of f_0 injective AND a U.I. where you apply this definition to some elements.

3.91. The hints for the previous exercise are relevant, but things are harder because the definition of surjectivity has an existence claim in it, so on top of everything else is a proof based on E.G. I promise that you will have to use E.I. from the definition of f_0 surjective on $\mathbf{R}$ to get a candidate for the thing you need for g_0.

Section 3.4.3

3.92. This is an old standard; the "trick" needed is to split off from the sum up through $n_0 + 1$ the last term, and then use the "induction hypothesis" on the sum of the rest of the terms.

3.93. The trick for this is virtually the same as that for the last one, although the algebra is different.

3.94. This one requires some more algebraic ingenuity; what is required is to cube out the right hand side *sensibly*, so that you have the term the induction hypothesis helps with, and then do some overestimation on some ugly looking terms left over.

3.95. After you do it, you might want to think of a neater way to cope with $P(1)$, $P(2)$, ..., $P(n)$ than this list.

3.96. No hint needed, but which kind of induction are you using?

3.97. Again no hint, and again what kind of induction are you using?

Section 3.4.4

3.98. Realize that there are two hypotheses on A and B if we agree not to write explicitly that they, and X, are sets.

3.99. You may be out of practice at using examples. Try some small functions and some real-valued functions, give some non-examples, and when you get to the special case of some χ_S, try some good small and extreme examples. It is perfectly fair to go back and look up examples you yourself constructed for χ_S some time ago, but get yourself completely refreshed on this stuff before going on.

3.100. This is the proverbial second chance.

3.101. Think of those big universal quantifiers out in front as letting you do something relevant without actually having to think.

3.102. What are the choices for the proof of an implication? Which would you choose here? Why?

3.103. No hint available, but this is just like previous exercises (which I guess is a hint, come to think of it).

3.104. You need an element (not one of the sets) to apply it to.

3.105. No hint, but please spend a solid amount of time at this. Here's the first time you've been asked to do some discovery, so invest in the process.

3.106. Try around Exercise 3.79.

3.107. A case with two subcases may not be pleasant, but you never would have gotten here without quantifier mechanics.

Section 3.4.5

3.108. No hint.

3.109. The last hypothesis is really an existence claim: for each s in S, there exists (something). Fill in the something. After the usual clearing away of universal quantifiers for what you want to prove, use this last hypothesis first. A small example to accompany the proof may help; after you have cleared away the universal quantifiers, and used this last hypothesis, draw a picture of what you know is in R. What do the other hypotheses guarantee is also around?

3.110. The proof really subdivides into three pieces. In each piece, you have a perfect chance to practice the art of stripping off universal quantifiers, getting the "fixed but arbitrary" thing or things, applying some other universals to it, and deducing what you want.

3.111. It may help to formulate the definition of injective (Definition 1.2.1) so that the quantifiers are unmistakable. Having dealt with them, you need to apply the two universals in your hypotheses; the thing to be careful about is that you may only apply some universal about f to elements of *the domain of f*, and similarly for g. A picture or small example where the elements of the three relevant sets all look different (say, 2, a, X) may be useful.

As a hint of last resort, here are the steps at the end of the proof you should have before you begin serious use of hypotheses at the beginning of the proof.

> $\cdots$
> Thus $x_1^0 = x_2^0$ by (?).
> Thus $(g_0 \circ f_0)(x_1^0) = (g_0 \circ f_0)(x_2^0) \Rightarrow x_1^0 = x_2^0$ by direct proof.
> So $\forall x_1, x_2((g_0 \circ f_0)(x_1) = (g_0 \circ f_0)(x_2) \Rightarrow x_1 = x_2)$ by U.G.
> Since this is the definition of injective, applied to $g_0 \circ f_0$,
> we have $g_0 \circ f_0$ injective.
> Since g_0 and f_0 were arbitrary, we are done by U.G.   ∎

3.112. The hints for 3.111 are useful again, suitably changed.

3.113. Pictures may again be useful. But things are different from Exercise 3.111 because the definition of sujective contains an existential quantifier; thus the proof includes along with everything else an argument founded on the E.G. form. Remember to work from the conclusion backward as far as possible! And two applications of E.I. to the definitions of f and g surjective will be needed.

3.114. The hints for the last exercise are relevant. After you are done, though, prove it by contraposition if you did not the first time, or some other way if you did use contraposition.

3.115. Those that are available are given above in Hints 2.20–2.29, 2.31–2.37, and 3.56– 3.72.

3.116. No simple "proof-structure" hint for this is possible.

Lab I: Sets by Example

4.1. The simplest example and non-example will do.

4.2. Give some non-numerical sets, too.

4.3. Give some example of the same set (by the definition of "=") with at least four apparently different descriptions in terms of orderings and repetitions.

4.4. No hint should be needed.

4.5. A large collection of examples is appropriate, including some in which the letter "x" does not appear. Can you find some non-examples in that the "condition" doesn't actually allow you to determine whether x is in the set or not? This would mean that in fact you didn't really have a statement when values for x were inserted. One way to do this is to have some other variables in your condition.

4.6. Full exploration, please. For example, did you find a pair of sets A and B such that their intersection is the same as their union? Smaller than their union? Larger than their union? Did you include any infinite sets? Find any whose intersection has no elements?

4.7. The object "1" is an element of $\{\underline{1}, \{1\}\}$, where we have indicated by underlining the element of the set that makes the assertion true. The object "$\{1\}$" is also an element of $\{1, \underline{\{1\}}\}$, where again we use underlining. The object "$\{1\}$" is a *subset* of $\{\underline{1}, \{1\}\}$, where the underlining again indicates why. Continue.

4.8. How many distinct areas should there be? Well, something could be in all of the sets (that's one area), in three but not four (that's four more, one for each of the four cases corresponding to the omitted set), in exactly two of the sets (that can happen six different ways) .... Don't forget "none of the sets," although the area outside in your picture takes care of that. Are there enough areas in your picture? Do they correspond to the right things?

4.9. No hint should be required.

4.10. Your pictures should have various areas in them. One thing to help ensure good exploration is to label each of the areas (for example, one of them corresponds to $A \cap B$). Another way is to try to figure out when each of them would actually be absent (that is, there are no elements corresponding to that part of the picture).

4.11. No hint is available — about all you can do is try different combinations of sets involving a and b.

4.12. The reason care is needed is in coping with the case in which you are using $\{\{a\}, \{a, b\}\} \subseteq \{\{c\}, \{c, d\}\}$. You may deduce from this that $\{a\} \in \{\{c\}, \{c, d\}\}$, and thus $\{a\} = \{c\}$ or $\{a\} = \{c, d\}$. It is tempting, but wrong, to discard the second immediately; remember that a set listing may contain repetitions, and so although it looks as if the set $\{c, d\}$ has two elements, and so can't be $\{a\}$, that need not be true if the listing happened to contain repetitions. But what if it did?

4.13. No hint should be required, but it helps to make the elements of A and B different to start with. What happens if they aren't, or if B is the same set as A?

Lab I: Exercises

4.14. Use Venn diagrams (along with other things) and realize that this is a name for a piece of a Venn diagram you didn't have a name for before.

4.15. The use of Venn diagrams helps. The custom is to make U a rectangle with its interior unshaded, with all the other sets as portions of the interior in the usual Venn diagram style. Can you prove any of your conjectures?

4.16. If you didn't try Venn diagrams, you simply haven't been paying attention. But there is more to do: what, for example, is $(A \cup B) \Delta B$? What about other "mixtures" of set operations including Δ?

4.17. One way to do it is to produce some complicated set whose elements are sets whose elements in turn are a, b, and c in some combinations. But there's another way: suppose you made an ordered pair whose first term was (a, b) and whose second term was c? Does this work? Prove it.

4.18. Venn diagrams help a lot and answer the question for two sets. For three sets, note that if you start with $\#A + \#B + \#C$, then everything that is in a pair of the sets has been counted twice, so you had better subtract off the number of things in all of the two-fold intersections. But there is then a piece of your Venn diagram for which you counted every element three times and then subtracted it off three times. Oops! Better add it in again . . . .

Lab II: Functions by Example

4.19. Very, very thorough exploration, please. For example, suppose A and B are the same set. What would the "identity function" look like as a set of ordered pairs? What would a constant function look like?

4.20. Luckily, you have already a large class of examples to try this out on. Also, it allows you to critique your collection from the previous problem: did any of your examples have range different from codomain?

4.21. No hint required.

4.22. If your examples included some familiar functions in their ordered pairs form, you ought to recapture them with this. If not, try the exercise of starting with a familiar function, turning it into its ordered pairs version, and then turning it back.

4.23. Can you say precisely how to go from a table to the ordered pairs version?

4.24. Of course, you can't give the complete listing of either, which is the whole point.

4.25. Anything involving f^2 is doomed; the point is that the "squaring" (or whatever) is something about what happens to *values*.

4.26. Practice carefully. An example worth noting is the Sine function, in which "Sine" is the name of the function, although we think of the formula f given by $f(x) = \sin x$. The subtlety here is that you really don't have a formula for the Sine (you do? what is $\sin 2$?).

4.27. Luckily, by now you have a fairly full collection of examples of functions in both ordered pair and functional notation. These definitions deserve exploration with all of them.

4.28. No hint is needed, although diligence is.

4.29. What is needed to have $(a, c) \in g \circ f$ is that there be $(a, b) \in f$ and $(b, c) \in g$ for some b. So there is certainly a picture, with two arrows, one from a to b and one from b to c, which could be drawn in the function picture discussed before 4.28, suitably modified to include the two functions f and g. Also, to have $(a, c) \in g \circ f$ is to have $(g \circ f)(a) = c$. But $c = g(b)$ since $(b, c) \in g$, and so ... . Thus we get back to the usual notation as expected.

4.30. Recall that the definition of function requires that each element in the domain occur as the first member of an ordered pair (the condition to ensure this is the easy one to miss). And that is another condition needed for a function; what condition on f ensures this condition on the "reversal" of f?

4.31. No hint available.

4.32. The identity function is easy, particularly if you rewrite it in the more familiar functional notation. And then to check that you have gotten the inverse, it merely comes down to composition of functions, one f, the other your candidate for f^{-1}.

Lab II: Exercises

4.33. Examples with about three elements per set are plenty; also, suppose $h(a_1) = h(a_2)$ for some a_1 and a_2, but $f(a_1)$ and $f(a_2)$ are distinct?

4.34. What if f happens to be surjective?

4.35. The only problem is the psychological hurdle of constructing an ordered pair one (or both) of whose members is itself an ordered pair.

4.36. Remember that everything eventually comes down to the level of values. For example, to find the identity element, you want some function id so that $f + id = f$. What does this become at the value level? What does this say about each value of id? Now assemble these values into a function.

4.37. Hint: one of the sets is f itself. Now, what other sorts of sets can sensibly be intersected with a collection of ordered pairs?

Lab III: Sets and Proof

4.38. Well, let's assume $S \in S$ and see if we find a contradiction. For this to be true, we must satisfy the condition in the definition of S, namely, $x \notin x$, with S substituted for x. This yields $S \notin S$, so the assumption $S \in S$ yields $S \notin S$, a contradiction. Continue.

Lab III: Exercises

4.39. Recall that '$\Leftrightarrow$' is 'if and only if,' and so some statement '$P \Leftrightarrow Q$' is the same as '$(P \Rightarrow Q)$ and $(Q \Rightarrow P)$.' This in hand, things are straightforward manipulation of universal quantifiers.

4.40. Argument by cases is useful for handling the "or" in the definition of union, after that definition is inserted in the appropriate place in the proof. And surely you were going to use the strategy of proof by two containments.

4.41. Straightforward; you do have to deal with what $x \notin (B \cup C)$ means. One approach is to think a little; the other approach is to formally negate $x \in (B \cup C)$.

4.42. Realize that this is (or can be) really four subproofs, since there is an if and only if, and since each of its parts might involve a double containment proof. Further hint: proof by contradiction.

4.43. There are several strategies for trying to do this sort of problem. One is to construct enough examples so that you are sure, in advance of trying to prove anything, of the truth or falsity of what you have. Another approach is to simply embark on trying to prove each of them; if you get stuck at some point where it just seems that your hypothesis couldn't possibly yield what you need, then you look for a counterexample based on the perceived discrepancy ("this all by itself couldn't possibly guarantee that, because of this example ... oh, there's my counterexample").

4.44. No hint available or needed.

4.45. If you didn't make up a good example collection, you deserve whatever happened to you. It is, for some people, possible to prove all these things on the purely formal level, but even if you can do that you shouldn't. A useful fact to remember for one of the problems is that $A \in \mathcal{P}(A)$. For the last statement, please make sure you distinguish between $\emptyset$ and $\{\emptyset\}$.

4.46. Recall that you have two ways to prove '$\forall n(P(n))$' if the relevant n are positive integers, although you may tend only to think of one.

Lab IV: Functions and Proof

4.47. "Each element of ..." clearly calls for $\forall$; less easy to see is that the relevant quantifier is $\exists$ for the remainder, but even with this hint things aren't too easy.

4.48. $\forall a(a \in A \Rightarrow \exists b(b \in B$ and what?$))$.

4.49. $\forall a((a \in A$ and $(a, b) \in f$ and $(a, c) \in f) \Rightarrow$ what?$)$ What is needed to ensure that the apparently different pairs (a, c) and (a, b) are really the same? By the definition of ordered pair, how could this happen?

4.50. $\forall f \forall a((\lim_{x \to a} f(x) = L$ and $\lim_{x \to a} f(x) = M) \Rightarrow$ what?$)$.

4.51. $\forall a \in A(\exists b((a,b) \in f))$ is one. Note, though, that we have hidden a quantifier in this language, which may make proving things harder, not easier.

4.52. This is worth doing but not worth a hint.

4.53. Surjective requires that something exist, injective that something is unique.

4.54. You've explored these with examples before. One new kind of exploration is to think about how, given the quantified definitions, proofs are likely to go. And you do have to check that f^{-1}, as defined, is a function.

Lab IV: Exercises

4.55. You may wonder why there is no requirement that f^{-1} be a function here, but it is there. Recall that we have only defined composition of functions, and so when we write things like $(f^{-1} \circ f)$ there is implicit that assumption that each is a function.

4.56. On one hand, you have that two sets of ordered pairs are equal, and want to show that two other conditions. It is useful to say, in terms of the set of ordered pairs comprising f, what is $domain(f)$? (The set of all x such that there exists a pair (x,b) ... .) The business of showing that for all x, $f(x) = g(x)$ is merely universal quantifiers and change to functional notation.

The other direction, in which you assume functional equality, is almost a reversal of the above arguments. Indeed, in some proofs, you will even read something like "the reverse direction follows essentially by reversal of the argument."

There is one subtle point. Which should be done first, some argument about equality of domains or some argument about how $f(x) = g(x)$ for all x? Answer: you won't really be able to deal with "$\forall x$" unless you already have the domains in control.

4.57. This is entirely diligence and good exploration. One way to get some examples (not a complete set, but useful for comparison) is to take some function example from the past and either add or delete ordered pairs. One possible example is to take the square function and add in all the negatives of the squares. Finally, as far as R^{-1} goes, in what regard does the definition of composition of functions really need to be changed to define composition of relations?

4.58. This is actually a little easier in the ordered pair formulation of functions, although you should do it both ways. The only tricky point, after the usual manipulations with quantifiers, is when you are confronted with $(g \circ f)(a_1) = (g \circ f)(a_2)$. The crucial thing to recall is that $(g \circ f)(x) = g(f(x))$. Now use the universal quantification on g injective, applied to the right thing, *which is not* x. What is the right thing? Note that it could hardly be x, since x is unlikely to be in the domain of g except by coincidence.

4.59. These are manipulation of definitions and quantifiers. You may fairly complain that the product of functions has not been defined, so what does $\chi_A \cdot \chi_B$ mean? This complaint is technically correct, but in Exercise 4.36 of Section 4.2.1 we defined addition of functions, so a little initiative is all that is required.

4.60. Remark that in a previous exercise (Exercise 4.35 of Section 4.2.1) you came up with what you thought was the right function. Now you have the definitions to prove it.

Lab V: Exercises

4.61. That this is important should be clear, since half of this book is about it. As for the results in the Proposition, it is simply a matter of quantifier and definition manipulation to get all of them.

4.62. The key to this is so easy you may have missed it: $f(C) = F(C)$ and $f^{-1}(C) = F_n(C)$. As for the last thing, realize that $C \triangle D = (C - D) \cup (D - C)$, and you have a host of results about how the function F and F_n behave on unions and set differences. Finally, you could prove things about $\triangle$ directly from the definitions, but it is reinventing the wheel to do so.

Lab VI: Exercises

4.63. One place to look is open intervals like $(-r, r)$; another is to intersect these intervals with the set of rationals, so as to get $\{s : |s| < r$ and s is rational$\}$. For the finite examples, three-set families work fine.

4.64. Exploration by examples is crucial, since these ideas aren't natural the first time you see them. For example, what is S_{x^2+x-4}?

4.65. If you think of an index as being a marking "tag" of some kind, to mark a set $S_\alpha \times T_\beta$ you need a tag with both the information in α and that in β. An analogy might be how you used to label positions in matrices. The other discoveries are easy explorations, and their proofs easy use of double containment for sets and quantifier manipulations. Also, think of these S_α and T_β as subsets of the real numbers $\mathbf{R}$, and draw pictures of the various sets of points.

4.66. One approach, although by no means the only one, is proof by contradiction. After all, what better way to prove that something is the empty set than to show that no thing can be in it? By the way, you ought to actually try to find such a collection of sets; open intervals again!

4.67. Half lines.

4.68. No hints needed.

4.69. An easy surjective function from S to S is I given by $I(x) = x$ for all x. Then $\mathcal{S} = \{S_S : S \in \mathcal{S}\}$.

4.70. There are a few special cases, I guess, if the family of sets is empty or has one element!

4.71. Done.

References

[1] George Pólya. *Mathematical Discovery: On Understanding, Learning, and Teaching Problem Solving (combined edition)*. John Wiley & Sons, Inc., New York, 1981.

[2] John Conway. *Subnormal Operators*. Pitman, Boston, Mass., 1981.

[3] Lewis Carroll. *Through the Looking Glass, and What Alice Found There*. Macmillan, New York, 1898.

[4] Martin Lewinter. *Graph Theory*. Monographs in Undergraduate Mathematics, Guildford College, Greensboro, NC, 1985.

[5] George Pólya. *How to Solve It*. Princeton University Press, Princeton, NJ, 1945.

[6] Uri Leron. Heuristic presentations: the role of structuring. *For the Learning of Mathematics*, 5(3):7–13, 1985.

[7] Alan H. Schoenfeld. *Mathematical Problem Solving*. Academic Press, San Diego, 1985.

[8] Philip J. Davis and Reuben Hersh. *The Mathematical Experience*. Birkhäuser, Boston, 1980.

[9] Friedrich Waismann. *Introduction to Mathematical Thinking*. Frederick Ungar, New York, 1951.

[10] Douglas R. Hofstadter. *Gödel, Escher, Bach: an Eternal Golden Braid*. Basic Books, New York, 1979.

[11] Ernest Nagel and James R. Newman. *Gödel's Proof.* New York University Press, New York, 1958.

[12] Paul Halmos. *Naive Set Theory.* Van Nostrand Reinhold, New York, 1960.

[13] David Tall and Shlomo Vinner. Concept image and concept definition in mathematics with particular reference to limits and continuity. *Educational Studies in Mathematics*, 12:151–169, 1981.

[14] Tommy Dreyfus. Advanced mathematical thinking. In Pearla Nesher and Jeremy Kilpatrick, editors, *Mathematics and Cognition: A Research Synthesis by the International Group for the Psychology of Mathematics Education*, ICMI Study Series, pages 113–134. Cambridge University Press, Cambridge, 1990.

[15] M. T. H. Chi, P. J. Feltovich, and R. Glaser. Categorization and representation of physics problems by experts and novices. *Cognitive Science*, 5:121–152, 1981.

[16] Brian H. Ross. This is like that: the use of earlier problems and the separation of similarity effects. *Journal of Experimental Psychology: Learning, Memory and Cognition*, 13(4):629–639, 1987.

Index

Undergraduate Texts in Mathematics

(continued from page ii)

LeCuyer: College Mathematics with APL.

Lidl/Pilz: Applied Abstract Algebra.

Macki-Strauss: Introduction to Optimal Control Theory.

Malitz: Introduction to Mathematical Logic.

Marsden/Weinstein: Calculus I, II, III. Second edition.

Martin: The Foundations of Geometry and the Non-Euclidean Plane.

Martin: Transformation Geometry: An Introduction to Symmetry.

Millman/Parker: Geometry: A Metric Approach with Models. Second edition.

Moschovakis: Notes on Set Theory.

Owen: A First Course in the Mathematical Foundations of Thermodynamics.

Palka: An Introduction to Complex Function Theory.

Pedrick: A First Course in Analysis.

Peressini/Sullivan/Uhl: The Mathematics of Nonlinear Programming.

Prenowitz/Jantosciak: Join Geometries.

Priestley: Calculus: An Historical Approach.

Protter/Morrey: A First Course in Real Analysis. Second edition.

Protter/Morrey: Intermediate Calculus. Second edition.

Ross: Elementary Analysis: The Theory of Calculus.

Samuel: Projective Geometry. *Readings in Mathematics.*

Scharlau/Opolka: From Fermat to Minkowski.

Sigler: Algebra.

Silverman/Tate: Rational Points on Elliptic Curves.

Simmonds: A Brief on Tensor Analysis. Second edition.

Singer/Thorpe: Lecture Notes on Elementary Topology and Geometry.

Smith: Linear Algebra. Second edition.

Smith: Primer of Modern Analysis. Second edition.

Stanton/White: Constructive Combinatorics.

Stillwell: Elements of Algebra: Geometry, Numbers, Equations.

Stillwell: Mathematics and Its History.

Strayer: Linear Programming and Its Applications.

Thorpe: Elementary Topics in Differential Geometry.

Troutman: Variational Calculus and Optimal Control. Second edition.

Valenza: Linear Algebra: An Introduction to Abstract Mathematics.

Whyburn/Duda: Dynamic Topology.

Wilson: Much Ado About Calculus.